Springer Series in Surface Sciences

Volume 67

Series editors

Roberto Car, Department of Chemistry, Princeton University, Princeton, NJ, USA

Gerhard Ertl, Abt. Physikalische Chemie, Fritz-Haber-Institut der Max-Planck-Gesellschaft, Berlin, Germany

Hans-Joachim Freund, Fritz-Haber-Institut der Max-Planck-Gesellschaft, Berlin, Germany

Hans Lüth, Institut für Schicht- und Ionentechnik, Forschungszentrum Jülich GmbH, Jülich, Germany

Mario Agostino Rocca, Dipartimento di Fisica, Università degli Studi di Genova, Genova, Italy

This series covers the whole spectrum of surface sciences, including structure and dynamics of clean and adsorbate-covered surfaces, thin films, basic surface effects, analytical methods and also the physics and chemistry of interfaces. Written by leading researchers in the field, the books are intended primarily for researchers in academia and industry and for graduate students.

More information about this series at http://www.springer.com/series/409

Sergey Samarin · Oleg Artamonov
Jim Williams

Spin-Polarized Two-Electron Spectroscopy of Surfaces

 Springer

Sergey Samarin
Department of Physics
The University of Western Australia
Perth, WA, Australia

Jim Williams
Department of Physics
The University of Western Australia
Perth, WA, Australia

Oleg Artamonov
St. Petersburg State University
St. Petersburg, Russia

ISSN 0931-5195 ISSN 2198-4743 (electronic)
Springer Series in Surface Sciences
ISBN 978-3-030-13139-5 ISBN 978-3-030-00657-0 (eBook)
https://doi.org/10.1007/978-3-030-00657-0

This Springer imprint is published by the registered company Springer Nature Switzerland AG
The registered company address is: Gewerbestrasse 11, 6330 Cham, Switzerland

Preface

Many important properties of complex multi-particle systems such as bulk solids, thin films, clusters, and metamaterials are driven by inter-particle correlation. These structures form a basis for modern nano-technology, solid-state electronics, and new branches of solid-state physics such as photonics, spintronics, plasmonics, and magneto-plasmonics. They find applications ranging from storage media and smart materials to quantum computers and quantum communication. From that approach, the topic of electron correlations is a focus of modern physics and triggers the development of new experimental techniques for studying electron correlation. Two aspects of electron correlations: Coulomb interaction and exchange (i.e., the consequence of the fermionic nature of electrons), require adequate techniques for their experimental investigation. One of the recently developed techniques concerns the detection and analysis of correlated electron pairs ejected from a solid surface upon electron or photon impact. These correlated electron pairs carry information on the electron correlations inside the sample. If the spin of incident electron or ejected electron is resolved, it adds an additional dimension to the spectroscopy. In the last several years, the interest in spin-polarized electrons at surfaces experienced rejuvenation due to the observation of new phenomena such as Quantum Hall Effect and Spin Hall Effect and the discovery of new materials: topological insulators and graphene, where the electron spin, the spin-dependent interactions in solids, and observational geometry play an essential role. The development of the spin-polarized two-electron spectroscopy technique and its applications form the central theme of the book.

In general, this research monograph has two aims. Foremost is to indicate the leading edge of how spin-polarized electron experiments have revealed the asymmetries inherent in the structure and interactions in solids and their thin film surfaces. Secondly, it is hoped that the next generation of researchers will be inspired to apply and develop their methodology, techniques, and concepts for the continuing advancement of knowledge and understanding.

Our earlier research was inspired to explore and pursue Fano's (1959) insight that "spin–orbit coupling" was "a weak force with conspicuous effects." Inescapably, electron exchange and spin polarization of all interacting particles led to studies of atoms, molecules, and solid-state matter.

The use of incident spin-polarized electrons to probe surfaces and thin films and their interfaces explores the scope offered by selected atoms, their combinations in layers on surfaces or as dopants to modify structure, dynamics, and function. The studies are inspired by the development of increased sensitivity of instrumentation, broad conceptual theory, and potential applications in social and scientific needs, particularly of magnetic layers and multilayered structures.

The effects of symmetry are usually available with carefully planned experimental geometries; for example, (i) normal electron incidence on a rotatable crystal surface, (ii) spin polarization directions within mirror symmetry planes of the surface, and (iii) rotation and interchange of detectors with respect to the surface normal.

The concepts of space and time have become more important in seeking relationships between physical observables and the spin–orbit interaction. The energy and momentum dispersion of spin-dependent scattering phase shifts and their time development have reached atto- and femtosecond consequences when assisted with fast and powerful lasers. The influence of the interfacial extent of spin–orbit effects in multilayered structures, of chiral and angular momentum effects of surfaces on the spin polarization of electrons, continues to expand.

The potential applications of fundamental spin scattering include low energy electrons, using atomically clean two-dimensional materials, where understanding how electrons interact with each other and how they move and scatter determine the functionality of devices.

Challenges for the future arise from the combined influence of electron correlation and spin–orbit coupling with emergent quantum phases and transitions in heavy transition metal compounds with 4D and 5D elements; then possible spin–orbital entanglement may modify the electronic and magnetic structure.

The authors thank many colleagues for ideas and encouragements through their published contributions as recorded in the references and interactions at conferences and laboratory visits, in particular: J. Kirschner, F. O. Schumann, J. Berakdar, H. Schwabe, R. Feder, G. Stefani, A. Morozov, S. Iacobucci, A. Ruoccoa, R. Camilloni, F. Offi, H. Gollisch, F. Giebels, N. Fominykh, O. Kidun, K. Kouzakov, T. Scheunemann, A. D. Sergeant, R. Stamps, D. Meinert, N. v. Schwartzenberg, C. Winkler, H. Schmidt-Böcking, M. Hattass, A. Ernst, P. Bruno, E. Weigold, M. Vos, A. S. Kheifets, I. E. McCarthy, C. Tusche, C. H. Li, Xiao Yi.

It is noted that active research, both experimental and theoretical related to the content of this book, has been stimulated continuously since 1990 by Prof. J. Kirschner.

The tutorial contributions from books are especially acknowledged.

U. Fano and L. Fano (1959), Basic Physics of Atoms and Molecules. Wiley, New York

J. Kessler (1985), Polarized Electrons, 2nd Ed., Springer, Germany

R. D. Cowan (1982), The Theory of Atomic Structure and Spectra. University of California Press, New York

C. Kittel (2005) Introduction to Solid State Physics 8th Ed., Wiley, New York

R. Feder (1985) Polarized Electrons in Surface Physics, World Scientific, Singapore

J. Berakdar (2000) Electronic Correlation Mapping From Finite to Extended Systems. Wiley, New York

E. Weigold and I. E. McCarthy (1999), Electron Momentum Spectroscopy. Kluwer Academic/Plenum Publishers, The Netherlands

H. Kleinpoppen and J. F. Williams (1980) Coherence and Correlation in Atomic Collisions. Plenum Press, New York and London

J. Kirschner (1985), Polarized Electrons at Surfaces. Springer, Germany

H. Ibach and D. L. Mills (1982) Electron Energy Loss Spectroscopy and Surface Vibrations. Academic Press, USA

J. S. Townsend (2000) A Modern Approach to Quantum Mechanics. University Science Books, USA

Perth, Australia   Sergey Samarin
St. Petersburg, Russia   Oleg Artamonov
Perth, Australia   Jim Williams

Contents

Chapter 1
Introduction

Abstract The concept of a time-correlated (coincidence) electron detection is introduced. It was demonstrated that two electrons generated by a single incident electron from a solid surface can be detected in a back-reflection geometry.

Two-electron coincidence spectroscopy, hereafter referred to as (e,2e) spectroscopy, provides important information on the details of the electron-electron interaction, the mechanism of the secondary electron emission from solids, and the dynamics of electron scattering by solids. The pursuit of the concept of the time-correlated electron detection and its feasibility continues to reveal much fascinating physics.

Time-Correlated Electron Pair Detection

Two-electron coincidence spectroscopy is based on the ability to detect two electrons generated by one incident electron from a surface.

These two electrons escape from the surface simultaneously, but arrive at the detectors with a time difference determined by their different energies, hence by their time of flight difference. This time difference for electron energies of the order of 10 eV and flight distance about 10 cm is about $10 \div 50$ ns with limits determined by the lowest energy of electrons that can be detected as determined by available instrumentation. This ability to detect time correlated electron pairs emitted from a surface was first demonstrated experimentally by (Kirschner et al. 1992). The following illustrative discussion proceeds with an exploratory approach of considering that surface of a sample is irradiated by electrons with a given energy of the order of 20 eV and with a given intensity of the electron beam of $I = N \text{ s}^{-1}$, i.e. N electrons per second.

Two electron detectors D1 and D2 are indicated (Fig. 1.1) with electronics that allow measurements of the distribution of the difference time between detection of two electrons by detectors D1 and D2. Now assume that there are no events when two electrons escaping from the surface are generated by a single incident electron. Then in the difference time distribution there will be only one peak located at the time difference equal to the average time between two successive incident electrons.

© Springer Nature Switzerland AG 2018
S. Samarin et al., *Spin-Polarized Two-Electron Spectroscopy of Surfaces*, Springer
Series in Surface Sciences 67, https://doi.org/10.1007/978-3-030-00657-0_1

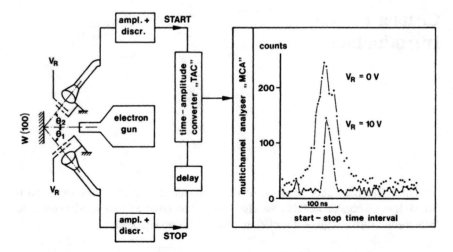

Fig. 1.1 Experimental set-up for the detection of time correlated electron pairs excited by incident electrons from a solid surface (left panel); detection time difference distribution of correlated electron pairs (right panel) Reprinted figure with permission from Kirschner et al. (1992). Copyright (1992) APS

Each of these electrons generates a secondary electron, and two of them may be detected by two detectors. If the incident electrons are distributed randomly in time, then the number of electron pairs detected by detectors D1 and D2 and separated by the time Δt will be distributed according to Poisson statistics with the maximum at $\Delta t = 1/NN$s, where N is the average number of electrons arriving to the sample per second. For example, for the incident electron intensity of $I = 10^5$ s^{-1} the position of the maximum will be at $\Delta t = 10$ μs. This maximum corresponds to the events where two detected electrons originate from two different primary electrons.

Now assume that there are events where two escaping from the surface electrons result from one incident electron scattered from the surface. These two electrons are detected singly, with one in each of the two detectors, with the time difference in the range of 0–50 ns because of the energy difference between the two electrons. Since faster and slower electrons can be detected by either of the detectors the distribution will be symmetric. An electronic time delay is introduced in one of the detection circuits to allow the full maximum time to be recorded. Thus, if the time difference distribution in a wide range of time (from 0 to more than 10 μs) is measured, there would be two maxima in the distribution: one at the time difference $\Delta t = 10$ μs ("accidental coincidences") and the second at the time difference of about 20–30 ns (time-correlated electrons or "true coincidence"). A tail of accidental coincidences extends to the region of "true coincidences" and forms a background. It is apparent that the background becomes smaller when the intensity of the incident electron beam becomes smaller. The two above-mentioned peaks will move apart from each other.

In the measurements the time-difference distribution is zoomed in the region of interest, i.e. in the region of "true coincidence" that is shown in Fig. 1.1 right panel. The time resolution in those measurements was about 5 ns. The broadening is due to the differences in flight time between fast and slow electrons which is demonstrated in Fig. 1.1 (right panel) by the different widths of the distribution obtained for two different retarding potentials on the grids. Increasing the retarding voltage suppresses the slow electrons, hence decreasing the range of possible transit times between fast and slow correlated electrons. The maximum of the peak corresponds to electrons with equal time of flight (TOF) in the two channels, hence equal energy. The tail on the right-hand side corresponds to fast electrons in the start channel and slow electrons in the stop channel, and vice versa for the tail on the left-hand side. The very existence of the peaks in Fig. 1.1 shows the existence of correlated two-electron emission events. The substantial decrease of the coincidence rate at high retarding potential points out the important role of low energy electrons in two-electron emission events. The fact that the true coincidence distribution curves are located at the same point on the time scale and are single peaked indicates that most of the electrons producing coincidences have similar energy. These results prove the existence of correlated electron pairs emission from a surface under electron impact and open the avenue for the development of the technique for studying such phenomena.

If the spin state of the incident or scattered electron is resolved, then the spin-dependent effects in the electron-solid interaction become observable. It is difficult to overestimate the above-mentioned information. From a fundamental point of view, it helps our understanding of the influence of spin-dependent interactions on the electronic structure of solids, as well as to elucidate a new class of phenomena termed "topological quantum phenomena" involving electron spin, the geometric (Berry) phase and special materials like topological insulators and graphene (Xiao et al. 2010; Shen 2004; Chiu et al. 2016; Moore 2010; Meunier et al. 2016). Potentially this technique enables tackling the problem of electron pair entanglement (Ghirardi and Luca 2004; Lednický and Lyuboshitz 2001) which continues to be dominant in understanding the deeper quantum nature of electron interactions. From the view point of applications, any information that can be provided by spin-polarized single- and two-electron spectroscopies on creation, control, transport and detection of spin states is relevant to the new promising field in solid state electronics, which is called "spintronics" (Soumyanarayanan et al. 2016).

The results presented in this book concern the development of the two-electron correlation spectroscopy of surfaces and its application for studying electron scattering dynamics as well as electronic structure of magnetic and non-magnetic surfaces, thin films and nanostructures.

Chapter 2
New Experimental Technique for Studying Electron-Electron Interaction, Electron Correlation, Mechanism of Electron Emission and Electronic Properties of Surfaces

Abstract A spectrometer for two-electron coincidence spectroscopy of solid surfaces in back-reflection geometry ((e,2e) spectroscopy) is described. The combination of the coincidence technique with time-of-flight electron energy measurements enables implementation of a "complete scattering experiment" on a solid surface, where the momenta of the incident and two scattered electrons are determined. The contributions of single-step and multi-step scattering events to the emission of two time-correlated electrons can be separated on the basis of experimental geometry and kinematics. Examples of applications of (e,2e) scattering to the study of various surface phenomena and electron scattering dynamics are presented.

The design of a tool for studying electronic correlations in a solid is based on the idea that two (or more) emitted electrons, resulting from the de-excitation of a solid target perturbed by an external energetic particle, carry information on their interaction (correlation) inside the solid. The ideas of (e,2e) had surfaced with Glassgold and Ialongo (1968) and confirmed with measurements by Amaldi et al. (1969) and Camilloni et al. (1972), and with gaseous atoms by Ehrhardt et al. (1969), Weigold et al. (1973) and Williams (1975). A detailed review of the application of the (e,2e) technique for studying the electronic structure of atoms and molecules and for the study of solids showed the way forward (McCarthy and Weigold 1976).

The essence of (e,2e) spectroscopy for atoms is the impact ionisation of a bound electron of an atom with measurements of the momenta of the two scattered and ejected electrons, so that energy and momentum of the previously bound electron can be calculated. Moreover, for high primary electron energies (~1,000 eV) the cross-section for such a reaction is proportional to the square of the bound electron wave function in a momentum representation. From that approach the technique provides access to the electron wave function in an atom or molecule. Similar information about solid surfaces motivated further studies.

Theoretical aspects of the application of (e,2e) spectroscopy to crystalline solids and surfaces were developed almost half a century ago (Smirnov and Neudachin 1966; Neudachin et al. 1969; Levin et al. 1972; D'Andrea and Del Sole 1978) and

© Springer Nature Switzerland AG 2018

S. Samarin et al., *Spin-Polarized Two-Electron Spectroscopy of Surfaces*, Springer Series in Surface Sciences 67, https://doi.org/10.1007/978-3-030-00657-0_2

led to new directions (Camilloni et al. 1972; Kirschner et al. 1992, 1995; Iacobucci et al. 1995a, b).

Explorations of experimental feasibilities were demonstrated early in transmission geometry for carbon films (Amaldi et al. 1969; Camilloni et al. 1972). With good energy resolution and higher primary energy the (e,2e) experiments in transmission geometry were extended to metal solid films and oxides (Hayes et al. 1988; Cai et al. 1995) and reviewed (Dennison and Ritter 1996).

In order to apply the two-electron spectroscopy for studying solid surfaces of bulk crystals, measurements in back-reflection geometry and with low primary electron energy were suggested (Artamonov 1983). Those conditions overcame the poor energy resolution that was intrinsic to high energy primary electrons and expanded the range of samples beyond the free standing thin film (~100 Å) to surfaces and then to nano-structures on the surfaces of bulk materials. In that case, the experimental method detected correlated electron pairs in the back-reflected hemisphere and analysed their energies and momenta.

2.1 Spectrometer for Two-Electron Spectroscopy of Solid Surfaces

The interaction of electrons with a metal surface results in the emission of secondary electrons over a wide range of energies from zero up to the primary electron energy. By convention, these electrons may be divided into three groups: (i) elastically scattered electrons, i.e. with no energy loss; (ii) inelastically scattered electrons, i.e. electrons that have lost part of their energy in single-particle or collective excitations on the surface, and (iii) "true secondary" electrons that are generated in multi-step scattering processes.

Accordingly, three types of electron spectroscopy were developed from the analysis of emission angle, energy and spin state of these groups of electrons. Schematically, the energy distributions of the secondary electrons may be represented as in Fig. 2.1 and labeled A, B, and C and usually their characterizations have been explored separately.

As was expected, there might be a scattering path, where two electrons generated by a single incident electron escape the surface. This process is referred to as (e,2e) scattering and represented in Fig. 2.1, which shows transmission (Fig. 2.2a) and reflection (Fig. 2.2b) geometries.

The two-electron coincidence spectroscopy of surfaces originates from the (e,2e) spectroscopy of atoms and molecules (Fig. 2.3). In that case, the differential cross-section of the detection of two correlated electrons as a function of the momentum q is proportional to the square of the wave function of the bound electron in the momentum representation. That is the momentum density distribution can be determined from the geometry indicated in Fig. 2.3. This spectroscopy is often called Momentum Density Spectroscopy (MDS). Application of this technique to a solid

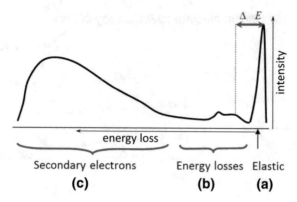

Fig. 2.1 Energy distribution of secondary electrons

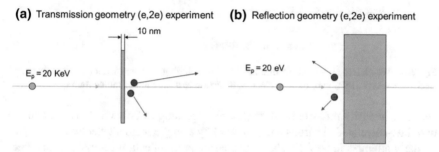

Fig. 2.2 a Transmission geometry (e,2e) scattering. **b** Reflection geometry (e,2e) scattering

target was first realized in transmission geometry (Camilloni et al. 1972) (Fig. 2.2a). In that case relatively high energy primary electrons (~20 keV) impinge onto a thin solid target (thickness ~10 nm) and two electrons (scattered and ejected) are detected in coincidence in a forward direction. The energies of both electrons are measured to provide information on the momentum density distribution in a thin film.

2.1.1 Coincidence (e,2e) Scattering from Surfaces in Back Reflection Geometry

In contrast to transmission through a thin film, and for an atomic target in a crossed beams geometry, the detection of two time-correlated electrons escaping from a solid surface is observed in a backwards direction relative to the incident electron momentum, as represented in Fig. 2.2b. An incident electron with energy of 20 eV (for example) impinges on a solid surface at a given incident angle, and two electrons generated by the incident electron escape the surface into the back hemisphere, where they are detected and analyzed. The high solid target electron density necessitates a

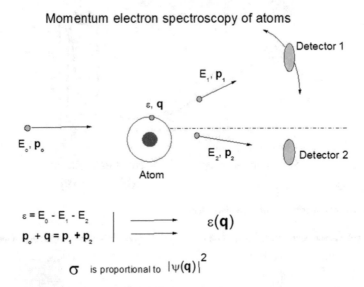

Fig. 2.3 Two-electron scattering from atoms. ε and $\mathbf{q}$—binding energy and momentum of the bound electron. σ—cross section of the two electron detection as a function of momentum $\mathbf{q}$

very low intensity incident beam to ensure, according to Poisson statistics, that the two time-correlated electrons are generated by a single incident electron.

Early attempts to perform such an experiment in reflection geometry using electrostatic energy analysers were not successful with low efficiency of the analysers (Avery 1977). The first successful two-electron coincidence experiment in reflection geometry on a solid surface was carried out in 1992 (Kirschner et al. 1992). It was convincingly demonstrated that two time-correlated electrons generated by a low-energy (10–50 eV) single incident electron could be detected in a back hemisphere and that the number of correlated electron pairs as a function of the primary electron energy could be measured. It was shown that the primary electron energy threshold of the excitation of correlated pairs depended on the work function of the surface and a simple kinematical model suggested the threshold (Kirschner et al. 1992). Since the energies of individual electrons of a correlated pair were not measured in this experiment, it was then difficult to prove that the correlated electron pairs were generated in single electron-electron collisions. Other successful measurements followed (Iacobucci et al. 1995a, b). A contribution from multi-step scattering events was difficult to exclude but became possible (Kirschner et al. 1995) only when the coincidence technique was combined with a Time-of-Flight (TOF) electron energy analysis. High efficiency position-sensitive detectors, based on large area (75 mm in diameter) micro-channel plates, were used for detection of correlated electron pairs and so measuring simultaneously the energy distribution of correlated electron pairs and their angular distribution within acceptance angles of the detectors (Kirschner et al. 1995; Artamonov et al. 1997). The next step toward "spin-polarized two-electron spectroscopy in reflection geometry" was the use of a spin-polarized electron beam in the experiment.

In the following, the experimental details of this technique that include the geometry of the experiment, the time-of-flight electron energy measurement, the combination of coincidence conditions with the time-of-flight energy analysis, data processing, and the spin-polarized electron beam are described.

2.1.2 Principle of Time-of-Flight (TOF) Energy Measurement

The principle of TOF measurement is simple (Uehara et al. 1990; Hemmers et al. 1998; Bachrach et al. 1975; Samarin et al. 2003; Kirschner et al. 2008; Wang et al. 2009). From the measured transit time T of an electron to traverse a distance L in a field free space between the sample and detector, the kinetic energy of this electron is:

$$E = (m/2)(L/T)^2, \tag{2.1}$$

where m is the electron mass. A reference point on the time scale is set by pulsing the incident electron beam. It follows from (2.1) that the relative energy resolution is

$$(\Delta E/E) = [(2\Delta T/T)^2 + (2\Delta L/L)^2]^{1/2} \tag{2.2}$$

and depends on the time resolution ΔT and the uncertainty of the flight distance ΔL. In the case of a point-like detector (small aperture channeltron, for example) and straight electron trajectories, the second term in (2.2) can be neglected. In turn, the time resolution is determined by the pulse width of the incident electron beam and the properties of the detector and electronics. However, the absolute energy resolution depends also on the electron energy to be measured and the flight distance L:

$$\Delta E = 2(m/2)^{-1/2}L^{-1}E^{3/2}\Delta T \tag{2.3}$$

As an example, for a pulse width of 500 ps, a flight distance of 0.1 m and electron energy of 10 eV, the energy resolution is about $\Delta E = 0.2$ eV.

With the distance between the centre of the detector and the sample equal to 126 mm, and an active detector area of 75 mm diameter, the acceptance angle of the detector is about 0.28 sr. In this case, the second term in (2.2) cannot be neglected because of the different flight distances for electrons arriving at different locations on the detector. The value of this term reaches 4% for the above example. To take into account the different flight distances for electrons arriving at different locations

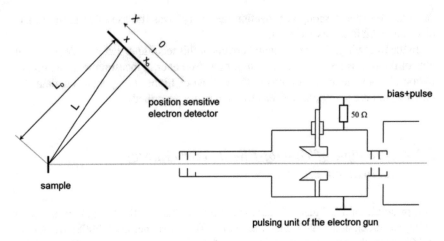

Fig. 2.4 Pulsing unit of the electron gun and a flight distance correction. (Reprinted from Samarin et al. (2003) with the permission of AIP Publishing)

on the detector, the position sensitivity of the detector is used. The procedure of the Time-of-Flight electron energy measurement and data processing is described below.

The electron source produces a pulsed electron beam with a pulse width of about 1 ns and repetition rate of 4×10^6 Hz. The pulsing unit is based on the deflection of the electron beam through a pair of deflection plates (Fig. 2.4). One of the plates is grounded and the other is connected to ground potential through a 50-Ohm resistor and permanently biased by a negative voltage to deflect the beam out of the exit aperture. A short positive pulse from a pulse generator (HP 8082A) arriving to the second plate "opens" the aperture to produce a pulse with a width of about 1 ns. The settings of the bias, pulse amplitude, pulse width and the focusing electrodes voltages are adjusted experimentally observing the width of the elastic maximum of scattered electrons on the time-of-flight scale. The bias voltage is in the range of 5–7 V and the pulse width and the amplitude are usually (2–5) ns and (4–5) V, respectively. The pulse applied to the deflection plate usually has a Gaussian (sometime triangular) shape. In the best case, the width of the time distribution of elastically scattered electrons was about 600 ps, but routine measurements are usually made at the elastic maximum width of about 1 ns. To apply a positive pulse on top of the negative bias to the deflection plate of the pulsing unit the circuit shown in Fig. 2.5 was used.

An alternative way to pulse the spin-polarized electron beam is to pulse the laser, which is used to generate electrons from a photocathode (Schumann 2010 PRL).

Every electron reaching the MCP may give rise to a fast pulse which is amplified and shaped by a constant fraction discriminator (CFD) whose output starts the Time-to-Amplitude Converter (TAC) which is stopped by a delayed (by about 200 ns) pulse from the pulse generator. In that way the amplitude of the TAC output pulse is proportional to the time difference between start and stop pulses and is recorded by an Amplitude-to-Digital Converter (ADC). The time window of the TAC is chosen to

Fig. 2.5 Circuit that
combines DC bias and pulse

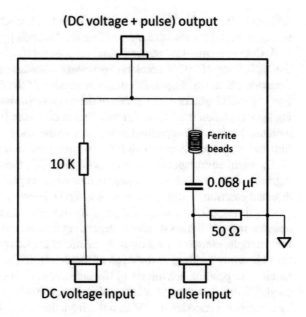

be 200 ns to cover an appropriate large range of electron energies and keep the pulse
repetition rate more than 4×10^6 Hz. For example, for a primary electron energy of
25 eV the lowest energy that can be detected in the chosen time window is 0.5 eV.
The time distance between generator pulses is set at 250 ns (i.e. the repetition rate
is 4×10^6 s^{-1}) that is 50 ns larger than the time window of TAC. In that way there
is a smaller probability of the slow electron "tail", generated by previous incident
electron, to be detected within a 200 ns TAC window.

In any clean surface science experiment, the two-electron coincidence measurements require UHV conditions with a base pressure in the 10^{-11}Torr range, or lower,
to minimise changes in the studied surface. To avoid an influence of magnetic field
on the electron trajectories, the residual magnetic field within the vacuum chamber
is reduced to less than about 5 mG using Helmholz coils. A sample under study is
mounted on a movable holder, kept at the ground potential and cleaned in vacuum
prior to measurements. All electrodes facing the sample are at the ground potential
to assist a field free space in the scattering region. The incident beam electron current
can be measured in a Faraday cup (FC) placed on the beam axis behind the sample
and then moving the sample off-axis. For a very small incident current, an MCP
detector can be used for an estimation of the incident current provided the efficiency
of the detector is known.

For detection of electrons, position-sensitive micro-channel plates (MCPs) with
a large detection area are most suitable. They possess very high detection efficiency
and allow effective simultaneous detection of the angular distribution of electrons
(within the acceptance angles of detectors) while the timing resolution of their resistive anodes is used for position-sensitive identification. For example, each of two

detectors, 75 mm in diameter, consists of two micro-channel plates in a Chevron arrangement with a resistive anode (Quantar Technology, Model 3394). A grounded grid (Copper mesh, 92% transmission) is mounted in front of the first MCP to allow the application of 200 V accelerating voltage between grid and MCP for increasing detection efficiency. Figure 2.6 shows schematic of the detector and voltage divider (upper panel) together with a photo of the detector (lower panel). The fast pulse timing signal is taken from an additional 30 mm diameter, 0.2 mm thick Titanium plate mounted behind, and separated from, the resistive anode by an insulating polyamide film and connected directly to a BNC vacuum feedthrough.

The main advantages of the time-of-flight (TOF) technique are its simplicity and high efficiency because all energies are measured "in parallel". Then the energy of the detected electron, irrespective of its energy, is determined by measuring the arrival time with respect to a reference point on the time scale. Electrons arriving at the detector travel different distances depending on the arrival position on the detector. For example, electrons landing at the centre of the detector travel a shorter distance than electrons incident at the edge of the detector. The flight distance corrections require the position sensitivity of the detector and, besides the fast timing signal, another four pulses from the corners of the resistive anode are amplified and processed by the position encoder. The SCA pulse from the TAC is used as a gate to trigger the output pulses from the position encoder. Then three pulses, representing the arrival time T and the (x, y) position on the detector are processed by the ADCs and stored in a list-mode file in the computer. An MPA-3 multi-parameter acquisition system (FAST ComTec) is used for the data collection.

Instead of the simplest TOF set-up with a field—free travel space and straight electron flight trajectories, more complicated configurations for the electron coincidence measurements have been based on spherical or elliptical electrostatic reflection mirrors (Wang et al. 2009; Davydov et al. 1999; Artamonov et al. 2003, 2001). Their main advantages are large acceptance angle and improved energy resolution. For example, it was shown that time-energy dispersion of a retarding spherical field in an electron mirror configuration is positive and it increases with the electron kinetic energy (Artamonov et al. 2003) due to an increasing penetration of electrons with high kinetic energy into the retarding field. For an electrostatic spherical mirror spectrometer, with inner and outer radii of 10 and 22 cm respectively, an energy resolution of about 0.5 eV/ns and an acceptance solid angle of about 2.2 sr for electrons of 75 eV kinetic energy can be expected. A position sensitive detector allows retrieving the energy and the emission angle on the basis of the measured time-of flight and detection position on the detector.

The 'reaction microscope', where the combination of electric and magnetic fields enabled an acceptance angle of 2π (Hattass et al. 2004), with a hexagonal delay line anode (Jagutzki et al. 2002), several electrons with position and time resolution with a very low dead time could be measured. By measuring time of flight and impact position the initial momenta can be reconstructed to show the complete six-dimensional two-electron momentum final state. This technique was used for studying dynamics of two-electron photoemission from C(111) (Hattass et al. 2008). The full solid angle coverage provided a comprehensive view of the details of the emission process

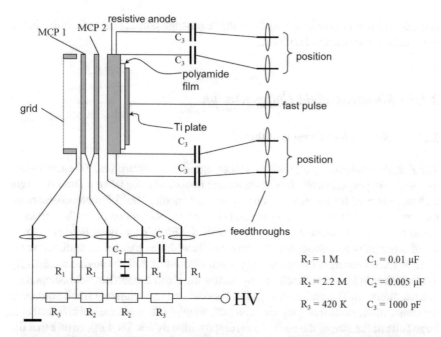

Fig. 2.6 Schematic of the MCP based position-sensitive detector (upper part). (Reprinted from Samarin et al. (2003) with the permission of AIP Publishing) and image of the detector (lower part)

and allowed for a quantitative study of the "exchange-correlation (*XC*) hole" in the two-electron momentum final state.

2.1.3 Magnetic Field Compensation

2.1.3.1 Static Field Compensation

The Earth's magnetic field provides a most significant concern for low-energy electron scattering experiments. It is an especially important issue when a Time-of Flight technique is used for the electron energy measurements. The ToF method described here requires field free space and straight electron trajectories on the path from sample to detectors. A common method to reduce the Earth's magnetic field is to use a set of three pairs of orthogonal Helmholtz coils with one pair for each direction (*x*, *y*, *z*). A fixed current is passed through each Helmholtz coil to provide a relatively gradient-free field which exactly compensates the Earth's magnetic field component in each direction. The region of zero-field is confined to a relatively small volume in the centre of the coils. For long electron path lengths and experimental convenience, large coils of the size of the entire room (approximately $6 \times 5 \times 4$ m), ensure that the maximum field free space with minimum field gradient space, is provided.

Another major concern is that all materials employed in the construction of the apparatus (including chamber materials, valves, bolts, washers), are non-magnetic. The metals from which the components are constructed must be carefully selected to ensure they are non-magnetic (notably near welds and nut threads) and remain so after baking or straining. Remnant and induced magnetism may be reduced by a process of degaussing, where a reasonable AC current is passed through a large coil around the vacuum chamber (about 1.5 m in diameter) to produce an alternating magnetic field which effectively takes the metals back and forth through their hysteresis loops. The magnitude of the current through the coil (and the field) is slowly reduced such that different (size and composition) domains end up in different orientations to decrease long range magnetic order. The positions and orientations of the coil, and the effectiveness of the procedure, may change after each cycle of degaussing and are dependent on the design and use of the apparatus.

2.1.3.2 Dynamic Field Correction

While static Helmholtz coils reduce the Earth's magnetic field they cannot compensate for the dynamic field changes present in a typical building. These include objects such as lifts in the building, or chairs, tools, gas cylinders being moved about in the area (not necessarily only within the lab). In fact most objects carry or create a small magnetic field that can impact negatively on a low-energy electron experiment.

For example, in the lab that was located not far from two major lifts, the incident beam current varied by two or three orders of magnitude, and the ratios of elastic to secondary electrons changed dramatically as the lift moved up or down between the basement and the fifth floor. Then a flux-gate magnetometer showed the lift produced a 10–20 mG field within the lab which could vary the full range in under a minute.

Two suitable methods exist to deal with this dynamic field problem. First, mu-metal with a high magnetic permeability and appropriate thickness, is used to construct a closed shell so that external magnetic fields will be contained within the material and will not enter the interior volume. This option is usually not suitable for large volumes or when heavy panels need to be moved to make simple procedures on the apparatus. The second method involves a feedback loop to dynamically measure and drive a magnetic coil to compensate any changes in real time. This method is described below.

To carry out the field compensation a new flux-gate magnetometer and high current bi-polar power supply were used. A computer equipped with eight 16-bit differential analogue-to-digital converters and two 16-bit digital-to-analogue outputs read the flux-gate magnetometer and drove the bi-polar power supply. The control program reads all three axis of the flux-gate magnetometer and uses one of them (user selectable) as the process variable. This variable is passed to the feedback loop, and a new output variable is calculated based on the historical changes in the variable with the first and second derivatives taken into account. The output variable is used to drive the bi-polar power supply which generates the compensating magnetic field.

To calibrate the system one magnetometer was attached outside and close to the main chamber scattering region and a second magnetometer was inserted at the sample position. The lifts were disabled at the basement of the building, that is the furthest and constant distance from the apparatus. The static Helmholtz coils were tuned such that the inserted magnetometer read zero field; thus compensating the Earth's magnetic field plus any other static fields. The readings on the outside magnetometer were recorded; the field outside the chamber was not zero as the probe was at a different location in the static fields, plus external objects may act to perturb the field slightly. The field values recorded by the external magnetometer probe represented the levels (of about 85 mG) that the dynamic system tries to maintain as they correspond to a "real" zero field in the scattering region.

The effectiveness of the final system of coils and apparatus is indicated by noting that when operating on one axis with one magnetometer the compensating program can carry out 20–30 field corrections per second. This easily nulls the movement of the lift, plus any "human speed" motions such as moving tools about the lab. Without compensation the magnetic field in the z-axis varied by about 10 mG and after compensation the field changes in the z-axis were reduced to a normal distribution about the ideal value with a standard deviation of approximately 0.05 mG.

2.1.4   Combination of Time-of-Flight and Coincidence Techniques

The geometric arrangement and electronics of the two-electron coincidence spectrometer are schematically shown in Fig. 2.7.

When an incident electron generates a correlated pair of electrons and they are detected by two detectors, two pulses from Constant Fraction Discriminators (CF) start two time-to-amplitude converters (TAC). A stop pulse to both TACs comes from a logic unit that delivers a stop pulse only when two delayed and shaped to 200 ns long square pulses by two independent gate generators (EG&G Ortec GG8010 Octal Gate Generators) from the detectors, and a delayed short trigger pulse from the generator, coincide (see Fig. 2.8). An output pulse of the TAC, with amplitude proportional to the time difference between start and stop pulses, is fed to the multi-channel analyzer and located into the channel corresponding to its amplitude, i.e. to the flight time, of the electron.

To minimize the dead time of the TACs the pulse generated by the coincidence unit can be used as a "start" pulse and two pulses from two detectors, after a proper delay, are used as "stop" pulses. This "reverses" the time-of-flight spectrum on the time scale (or multi-channel analyzer), so that the elastic peak resides at the largest channel number of the spectrum and slower electrons are represented by lower channel numbers. Usually the average incident current is in the 10^{-14} to 10^{-13} A range, and that implies, on average, less than one electron per incident pulse. This decreases

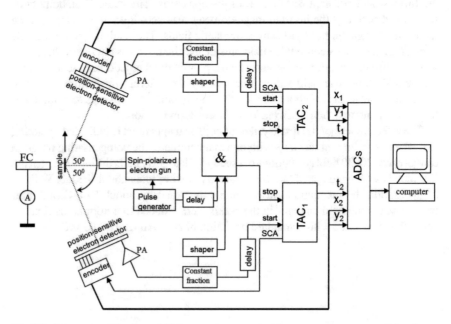

Fig. 2.7 Schematic of the time-of-flight two-electron coincidence experiment on a solid surface

Fig. 2.8 Time-to-amplitude electronics operating in coincidence mode. The detection of an electron starts a 200 ns square pulse. For a coincidence event to occur the square pulses from both detectors must overlap with the shaped 20 ns trigger pulse

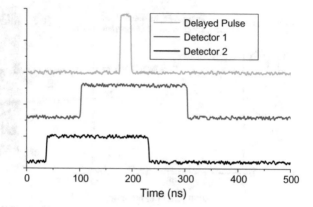

the number of accidental coincidences when two detected electrons are generated by two different primary electrons.

Besides the timing pulses, the electron arrival positions on the detectors were measured. The position sensitivity of the detectors allows the measurement "in parallel" of the angular distribution of the electrons. In addition, one can observe the electron diffraction patterns and estimate the electron beam size on the position-sensitive detectors. The data acquisition PC employs a FAST ComTec analogue-to-digital conversion (ADC) system (FAST MPA-3 with two Quad ADC 7074). The system has been configured to ensure all three signals (time, x, y) from each detector remain together. This is achieved by beginning conversion when a non-zero voltage signal is detected in either of the two TAC outputs. All six voltages, three from each detector, are converted to 12-bit digital signals. When operating in non-coincidence (singles) mode this method of conversion results in large numbers of null events, where only one detector has an electron (while the other has none). These events are easily handled in post-processing of the data sets and do not occur in coincidence mode. The 12-bit signals from the ADCs are read by the PC as 16-bit variables, where the last bit from each TAC channel contains the digital output state sent to the liquid crystal unit controlling the polarization of the incident beam. This method provides a record within the data stream of the spin-state of the incident electrons. The data are stored by the software MPA-3 in list-mode format allowing post-processing by external software. In that way the detected time-correlated electron pairs present a six-dimensional array (for each polarization of the incident beam), which can be projected onto any two-dimensional or three-dimensional distributions. For example, one can plot the number of pairs (coincidence events) as a function of E_1 and E_2, where E_1 and E_2 are energies of both correlated electrons (see Fig. 2.9).

It is worthwhile to note that the above-described two-electron coincidence spectrometer can be used for measuring low energy Electron Energy Loss Spectra by switching off the coincidence conditions in the electronic setup of Fig. 2.7.

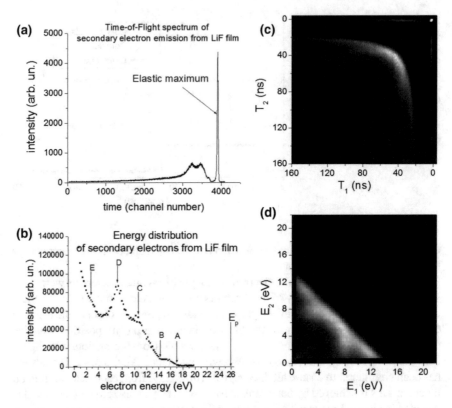

Fig. 2.9 a, c Time-of-Flight spectra of secondary electrons excited from LiF film by 26 eV primary electrons; **b, d** the same spectra but in energy representation

2.1.5 Data Processing and Various Presentations of Measured (e,2e) Distributions

As mentioned above, every measured spectrum is a six-dimensional array of data (for one spin orientation of the incident beam) that is: (x_1, y_1) coordinates of the detected electron of a pair on the first detector and (x_2, y_2) coordinates of the second detected electron of a pair on the second detector; t_1 and t_2 are arrival time of the first electron and the second electron respectively. Such an array of data can be projected onto two-dimensional or one-dimensional graphs. As an example, consider a transformation of the six-dimensional array of a measured (e,2e) spectrum into a two-dimensional energy distribution of the correlated electron-pairs.

To convert a time-of-flight distribution to an energy distribution, it is necessary to know the exact flight distance from the sample to the impact point in the detector plane for each electron trajectory and to calculate t_0—the time when the primary electron hit the sample and when scattered electrons leave the sample. For the electron detected at the time T and at the position (x, y) on the detector the flight time is $t = T - t_0$ and the flight distance L is (see Fig. 2.4):

$$L = \left(L_0^2 + x^2 + y^2\right)^{1/2}. \tag{2.4}$$

The flight time of the elastically scattered electrons with well-defined kinetic energy E_0 can be used to calculate the t_0 and it is also dependent on the detection position. To reduce this spread of TOF the measured TOF is scaled to the unique flight distance equal to L_0 (see Fig. 2.4). The sharp narrow maximum in the TOF distribution in Fig. 2.9a denotes the arrival time T_0 of the elastically reflected electron from the sample to the center of the detector. Using T_0, the distance between the sample and the detector center L_0, and the incident electron energy E_0, the t_0 is calculated using

$$t_0 = T_0 - L_0 C^{-1} E_0^{-1/2}, \quad \text{where } C = (2/m)^{1/2}.$$

Then the energy E of the detected electron can be calculated as follows:

$$E = L^2 (t \cdot C)^{-2} = \left(L_0^2 + x^2 + y^2\right)\left[(T - T_0)C + L_0 E_0^{-1/2}\right]^{-2}. \tag{2.5}$$

This calculation is completed using the time delay per consecutive channel in the spectra which is obtained from discrete delays applied to a pulse with the precise delay time measured on a high-speed oscilloscope. These reference delays are then applied to each arm of the spectrometer (left and right) to move the elastic peak (specular reflection of incident electrons into each detector) by a given number of channels. Knowing how many channels each delay moves the elastic peak the delay per TAC step in picoseconds/channel is determined. When this calibration is carried out for a number of delays ranging from 0.5 to 50 ns the results should be in reasonable agreement with each other. While transforming the time-of-flight distribution into energy distribution and building a corresponding histogram it is necessary to take into account the transformation of the time interval Δt to the energy interval ΔE. Figure 2.9c represents the two-dimensional time-of-flight distribution of correlated electron pairs. Each point of the distribution represents a correlated pair of electrons. One of the coordinates of the point shows the time of flight of the first electron of the pair from the sample to the detector. The second coordinate of the point represents time of flight of the second electron of the pair. Figure 2.9b, d show examples of the energy distributions of secondary electrons obtained from the TOF distributions (Fig. 2.9a, c using the above described procedure. Figure 2.9d represents the 2D energy distribution of correlated electron pairs. Each point of this distribution represents a correlated electron pair with the coordinates showing now energies of both electrons. Because coincidence experiments may take more than two weeks to yield reasonable statistics, it is essential that the processing program is able to

Fig. 2.10 Momentum-resolved (e,2e) spectroscopy

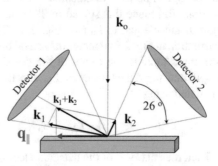

Position sensitive detection $\Longrightarrow$
momentum - resolved spectroscopy
(x_1, y_1, t_1) and (x_2, y_2, t_2) = correlated electron pair

$$\mathbf{k}_{1\parallel} + \mathbf{k}_{2\parallel} = \mathbf{q}_\parallel + \mathbf{k}_{0\parallel}$$

normal incidence $\Longrightarrow \mathbf{k}_{0\parallel} = 0$

$\mathbf{k}_{1\parallel} + \mathbf{k}_{2\parallel} = \mathbf{q}_\parallel$ - momentum of the valence electron

$(E_1 + E_2) - E_0 = \varepsilon$ - binding energy of the valence electron

combine statistical results which is achieved by reading and processing the list-mode data in blocks. The elastic peak is located for each block, of about 500,000 events, and used to convert the raw data into energy and position information. This method overcomes any shifts in the elastic maximum timing position between successive data blocks and experiments. The file reading and initial conversion is complete when all the files are combined into two large datasets, one for each spin orientation of incident electrons, containing $\{E_1, x_1, y_1, E_2, x_2, y_2\}$ events. When operating on coincidence experiments we read both singles and coincidence results to obtain four data sets; two spin-states for each singles and coincidence.

The use of position sensitive detectors allows measurements of the angular distributions of correlated electron-pairs and, consequently, to scan the component of electron momentum parallel to the surface. In the case of a single crystal sample, this component of electron momentum is conserved. Therefore, the measurement of the outgoing electron momenta enables the determination of the parallel component of the valence electron. Figure 2.10 shows how the momenta (wave vectors) of the incident electron $\mathbf{k}_0$ and the two outgoing electrons $\mathbf{k}_1, \mathbf{k}_2$ are related to the parallel component of the valence electron wave vector $\mathbf{q}\|$. In the experimental geometry shown in Fig. 2.10 the accessible range of $\mathbf{q}\|$ depends on the primary electron energy. For example, at 30 eV primary energy and normal incidence the accessible range of $\mathbf{q}\|$ is -1.5 Å^{-1} to 1.5 Å^{-1}. This range can be extended by increasing the primary electron energy or by using off-normal incidence.

Fig. 2.11 Two-dimensional energy distribution of correlated electron pairs excited by primary electrons with $E_p = 22$ eV from Co film at normal incidence. E_1—energy of the first electron, E_2—energy of the second electron of the pair

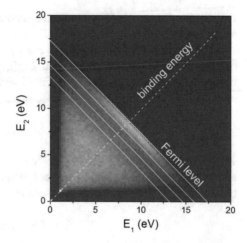

Beside true coincident events the (e,2e) spectrum usually contains accidental coincident events. These accidental coincidences may originate from noise pickup in one detector, two electrons entering the scattering chamber within one pulse, cross-talk between two detectors, and some other unexpected events. When considering coincidence data it is desirable to remove any accidental coincidences which contribute to a complex background. As accidental coincidences (involving detection of at least one electron) are principally single electron events (one real electron and one noncorrelated count) the background should resemble a non-coincidence spectrum from the same sample. A number of preliminary experiments have shown that this is the case. To determine (measure) the background the coincidence conditions in electronics are switched off and a dataset of singles is acquired before every coincidence experimental data acquisition run.

Before the background measurement can be used it must be normalised to the coincidence results. This normalisation is carried out at the region of the (e,2e) spectrum above the Fermi level line (see Fig. 2.11) but below elastic peak, i.e. upper corner of a two-dimensional energy distribution of the (e,2e) spectrum. This region is chosen for normalisation as there should be no real coincidence events in this region: electrons in the (e,2e) reaction cannot be excited from levels above Fermi level (strictly speaking this is correct at $T = 0$). This normalisation procedure is carried out for each spin orientation resulting in two normalisation coefficients. These coefficients are simply multiplied by the singles data which is then subtracted from the coincidence results.

After the time-of-flight dataset is converted to the energy distribution a few more presentations can be extracted from these distributions. For example, Fig. 2.11 presents a two-dimensional energy distribution of correlated electron pairs excited by incident electrons with energy of 22 eV from a cobalt film on W(110). Each point on this distribution represents a pair of electrons. One coordinate of the point shows the energy of the first electron and the second coordinate shows the energy of the second electron of the pair.

The Fermi level line separates two energy ranges: (i) the first one, where electrons can be excited from (below the Fermi level), and (ii) a second one, where electrons cannot be excited from (above the Fermi level). Energy conservation in the (e,2e) reaction indicates:

$$E_0 + E_1 = E_2 + E_3, \tag{2.6}$$

where E_0, E_1, E_2, and E_3 are the incident electron energy, valence electron energy, and energies of two outgoing electrons, respectively. It follows from (2.6) that the binding energy, with respect to the vacuum level, is determined by:

$$E_1 = (E_2 + E_3) - E_0 \tag{2.7}$$

For the total energy of pairs located on the "Fermi energy" line the binding energy is determined by:

$$E_1 = 17.5 \text{ eV} - 22 \text{ eV} = -4.5 \text{ eV} \tag{2.8}$$

This is the binding energy with respect to the vacuum level of electrons at the Fermi level, i.e. it is the work function of the Co surface.

The white lines on Fig. 2.11 separate three bands (in this particular example), in each of which the sum energy of the electron pairs is fixed (with uncertainty equal to the width of the band). For example, in the first band close to the Fermi level the total energy (or sum energy) of pairs is in the range of $E_{tot} = (16-7.5)$ eV, in the second band $E_{tot}=(14.5-16)$ eV and so on. The number of pairs within each of these bands as a function of the total energy assigned to the band shows the distribution of pairs along total energy, i.e. "total energy distribution". Figure 2.12 represents such a distribution.

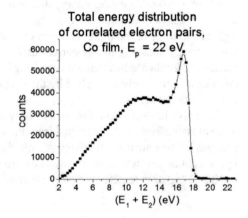

Fig. 2.12 Total energy distribution of correlated electron pairs excited by 22 eV primary electrons from a Co film at normal incidence and symmetric detection of electron pairs

Fig. 2.13 Energy sharing distribution of correlated electron pairs excited from Co film by 22 eV primary electrons at normal incidence and symmetric detection of electrons. $E_{tot} = (16-17)$ eV

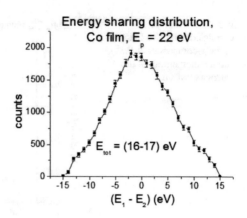

The onset of the distribution at about 17.5 eV indicates the position of the Fermi level on the total energy scale (this is true at zero temperature, but we accept this approximation given the finite energy resolution of the measurements). Indeed, the highest total energy of an electron pair corresponds to the excitation of the valence electron from the Fermi level. Obviously, electron pairs with maximum total energy are excited in single-step scattering events. Assuming that all pairs are created in single-step processes the total energy distribution can be easily transformed into the binding energy (with respect to the Fermi level, for example) spectrum by placing zero at the onset of the curve (17.5 eV in this case). The validity of this assumption is discussed below.

Within each of the bands of Fig. 2.11 one can analyse energy sharing between two electrons of pairs. Figure 2.13 shows one of such distributions for the total energy band $E_{tot} = (16-17)$ eV. At the point where $E_1 - E_2 = 0$ two electrons of a pair have equal energies. For this particular case, the maximum of the energy sharing distribution corresponds to the equal energy sharing between electrons of pairs.

Using parallel-to-the-surface momentum conservation one can plot number of pairs as a function of the k_{bx} component of the valence (bound) electron momentum:

$$k_{bx} = k_{2x} + k_{3x} - k_{0x}. \tag{2.9}$$

Figure 2.14 shows such a distribution.

The position of the detectors (strictly speaking, the position of the centres of the detectors) defines the scattering plane, i.e. the plane containing momenta of two outgoing (and incident) electrons provided the valence electron was at rest before the electron-electron collision. Since the valence electrons are moving inside the sample, the scattered (escaping from the surface) electrons are distributed in a range of angles (outside the scattering plane). However, using position sensitivity of detectors, one can plot k_x-distribution, i.e. the number of pairs as a function of only k_x component of $k_\parallel$.

Fig. 2.14 Number of correlated pairs as a function of the *x*-component of the surface momentum of the valence electron

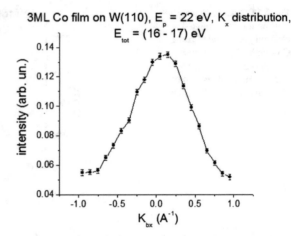

3ML Co film on W(110), E_p = 22 eV, K_x distribution, E_{tot} = (16 - 17) eV

2.1.6 Spin-Polarized Electron Beam

The formation of spin-polarized electron beam is based on photoemission from a GaAs crystal activated by a deposition of cesium and oxygen (Pierce et al. 1980; Kolac et al. 1988). A key-factor of obtaining polarized electron emission from this material is the spin-orbit splitting of the valence band of GaAs in the center of the Brillouin zone (Γ point) (Subashievy et al. 1998; Mamaev et al. 2008; Pierce et al. 1975; Pierce and Meier 1976; Erbudak and Reihl 1978; Reihl et al. 1979). At this point the degenerate p orbital is split into a fourfold degenerate $p_{3/2}$ level and a twofold degenerate $p_{1/2}$ level. The lowest state in the conduction band is a twofold degenerate $s_{1/2}$ orbital. The origin of the spin polarization generated by this material can be understood by considering the transitions between the m_j sublevels of the $p_{3/2}$ to the $s_{1/2}$ sublevels under the excitation by circularly polarized light. The selection rule dictates $\Delta m_j = \pm 1$ Considering the transitions from $p_{3/2}$ to $s_{1/2}$, say for positive helicity light σ^+, then, for a properly chosen wavelength (just equal to the band gap), there are two transitions allowed: (i) from $m_j = -3/2$ to $m_j = -1/2$, and (ii) from $m_j = -1/2$ to $m_j = +/2$ (see Fig. 2.15).

The first transition has a three times larger probability than the second. It means the excitation probabilities into the final state with spin projections $m_j = -1/2$ and $m_j = +1/2$ are not equal and there will be an imbalance between electrons with "spin-up" and "spin-down" in the conduction band. To allow these electrons to escape into the vacuum from the bottom of the conduction band the affinity of the surface must be reduced, for example by depositing Cs onto the surface in combination with oxygen adsorption.

Such activation leads to a negative affinity of the crystal surface, hence electrons from the bottom of the conduction band can escape into the vacuum by tunneling through a narrow barrier at the surface. The ultimate polarization P of the emitted electron beam is $P = (3 - 1)/(3 + 1) = 0.5$. In practice the polarization of the electron

600 °C for 20–30 min in the vacuum range of 10^{-10} Torr before the *Cs* deposition. The heating can be done by radiation from the filament or by electron bombardment of the rear side of the crystal holder. At the final stage of the crystal heating it is necessary to outgas the *Cs* dispenser by running through the dispenser a current of about (3.5–4 A) for about two minutes, and then letting the GaAs crystal cool down. When its temperature reaches 100 °C the *Cs* evaporation from the dispenser can be started by running a current of 3.5A through the dispenser. At this stage, the laser light with the wavelength of 830 nm should be switched on and aligned with the GaAs crystal while the photoemission current is monitored, for example by an Electrometer (e.g. Keithley 610CR). The vacuum in the spin source chamber must be in the 10^{-11} to 10^{-10} Tor range during *Cs* deposition. Usually, in about 10 min of the *Cs* deposition, a photocurrent of the order of a few nA appears. At this point oxygen should be introduced at a pressure of about 5×10^{-10} Tor. The photoemission current will increase and the *Cs* dispenser current and the oxygen pressure should be optimized by monitoring the photoemission current from the GaAs crystal. When the photoemission current stops increasing, and reaches saturation the activation procedure can be stopped. Usually, before the oxygen flow stops, some excess oxygen flow continues and the *Cs* dispenser current is kept at about (2.4) A. The continuous low-rate deposition of *Cs* when adjusted correctly can keep the emission current of the GaAs photocathode almost constant for a long time, up to some months. To avoid an influence of the magnetic field of the current through the *Cs* dispensers on the electron spin orientation a special holder of the current leads was designed (see Fig. 2.17).

Two *Cs* dispensers are mounted on a Ta plate within two slits in such a way that the current through the dispenser travels along two loops in opposite directions (dotted arrows on the Fig. 2.17), i.e. magnetic field of each of the loops is substantially compensated by the field of the other one. As checked experimentally the current through *Cs* dispenser does not influence the polarization of the electron beam. The *Ta* plate supporting the *Cs* dispensers is the first electrode of the electron optics of the gun. It is kept at a low potential in order to minimize the electron current (and electron energy) of electrons arriving on the electrode and, consequently, to reduce an electron stimulated desorption from the electrode, which, otherwise, would contaminate the photocathode and decrease electron emission. The longevity of the photocathode and stability of the photocurrent are essentially dependent on the cleanliness of the photocathode surface, i.e. on the vacuum in the spin-polarized source chamber. Hence, the clean vacuum in the (low) 10^{-11} Tor range is the prerequisite for the reliable operation of the spin-polarized electron source.

The circularly polarized light is produced using a semiconductor laser diode, controlled for example by a Schäfter-Kirchoff *SK 9733 C* power supply, linear polarizer and Liquid Crystal Variable Retarder (*LCVR*) with a Meadowlark Optics basic liquid crystal controller (*B1020*). To set voltages V_1 and V_2 of the *B1020* that are supplied to the *LCVR* for producing right and left circularly polarized light, a special rotating polarizer (polarization filter) and photo-detector are used (see Fig. 2.18). The laser light passes through the linear polarizer, which is oriented such that its axis makes $45°$ relative to the main axis of the *LCVR*, and then passes through the rotating linear

(a)

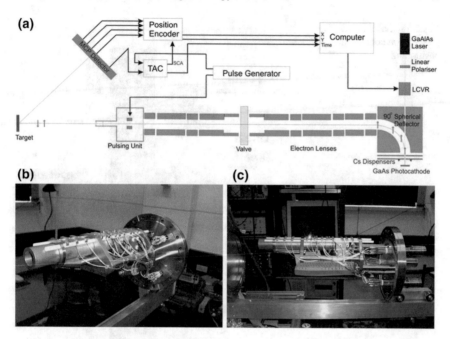

(b)   **(c)**

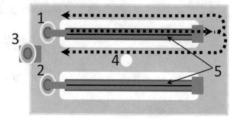

Fig. 2.16 **a** Schematic of the spin-polarized electron gun and one of the two detectors; **b, c** images of electron optics of the electron gun

Fig. 2.17 Two *Cs* dispensers mounted on *Ta* plate. 1 and 2—connections to the *Cs* dispensers (5); 3—connection to *Ta* plate

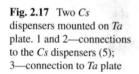

polarizer. This polarizer is connected to a turbine driven by compressed air to rotate with a speed of about 50 turns/s. At zero voltage on the *LCVR* the linearly polarized light passing through the rotating polarizer will have its intensity modulated as $Y = Y_0 |Cos(\omega t)|^2$, where ω is angular velocity of the rotating polarizer and t is time. The intensity $Y = Y_0 |Cos(\omega t)|^2$ is measured by the photo-detector and monitored using an oscilloscope. When a voltage is applied to the LCVR and makes a phase shift between two components of the **E** vector, the laser light becomes elliptically polarized and the amplitude Y_0 becomes smaller.

Ultimately, when the voltage on the *LCVR* is chosen such that the light passing through the LCVR becomes circularly polarized, the intensity Y becomes constant because it does not depend now on the angular position of the rotating polarizer. This voltage V_1 is set for the right-hand circular polarization of the laser beam. In a

Fig. 2.18 Laser beam
polarization control and
positioning

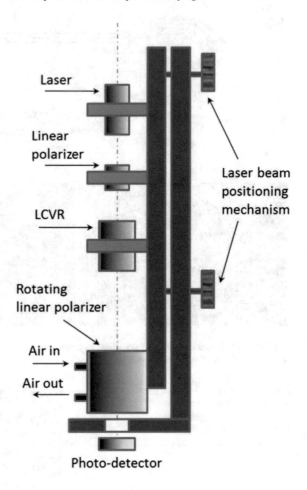

similar way, the second voltage V_2 is chosen, which produces the phase difference
of $3/2\,\pi$ and left-hand circular polarization of the laser light. The V_1 and V_2 on the
LCVR are controlled by a signal from the acquisition system. An external voltage
from the acquisition system (*MPA3*) changes the voltage from V_1 to V_2, i.e. changes
the phase shift between two components of the **E** vector of the laser light and convert
the right-hand circularly polarized light into the left-hand circularly polarized light
and, by consequence, changes the polarization vector of the electron beam from
"spin-up" to "spin-down". The trajectories of electrons emitted from the strain GaAs
crystal in the direction of the surface normal are then rotated by 90° using a quasi-
spherical electrostatic deflector. Since the electric field of the deflector does not
rotate the electron spin, the longitudinally polarized electron beam is converted to
the transversely polarized electron beam that is guided by the electron optics to the
scattering chamber. The degree of polarization of the electron beam depends on the

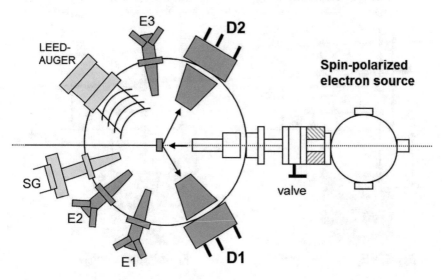

Fig. 2.19 Schematic view of the spin-polarized (e,2e) spectrometer that includes Main chamber and Spin-polarized electron source chamber. D1 and D2—electron detectors; E_1, E_2, E_3—OMICRON—type deposition guns for thin film deposition; SG—sputtering (ion) gun

properties of the photocathode, preparation of the surface, and the matching of the photon energy of the laser light to the band gap of the photocathode.

The scattering chamber and the spin-polarized electron source chamber are separated by a UHV valve (see Fig. 2.19) which allows venting separately either the scattering chamber or the spin-source chamber while maintaining the vacuum in the other chamber. An electrode of the electron optics at the position of the valve is split into two parts and the valve fitted in the gap between them such that both parts of the electrode have the same potential (of about 410 V) (see Fig. 2.16a).

2.2 Single-Step Versus Multi-step Electron Scattering in the (e,2e) Reaction on Surfaces

In the case of two electrons ejection in response to the electron impact on the solid surface one can imagine a number of possible pathways of the electron scattering process. Consider the energy relations between scattering electrons such that the energy balance of the (e,2e) reaction on a surface is represented in Fig. 2.20. There could be, at least, two pathways resulting in the same combination of energies of two outgoing electrons: single-step and two-step (multi-step) electron-electron scattering. It is clear that only in the case of single-step scattering event the final energies and

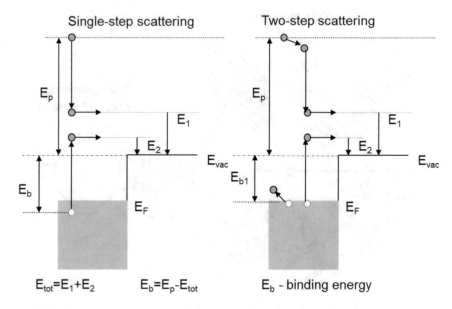

Fig. 2.20 Energy diagram illustrating single-step and two-step electron scattering

momenta of detected electrons determine the binding energy and momentum of the valence electron. Is it possible to distinguish experimentally single- and multi-step scattering paths? Yes, as will be demonstrated below.

Let us analyze possible pathways of the (e,2e) reaction on the energy diagram. From Fig. 2.20 it is seen that there are two different paths to the same final state with electron energies E_1 and E_2, when an incident electron with the energy E_p excites a valence electron of the target. Indeed, in a single-step process (left panel) an incident electron with the primary energy E_p transfers part of its energy to a valence electron with the binding energy E_b. Then both electrons being above the vacuum level can escape from the solid having energies E_1 and E_2. The second path includes, first, an inelastic scattering of the primary (incident) electron, which transfers a small fraction of its energy to a valence electron. The valence electron is excited above the Fermi level but it is still below the vacuum level and cannot leave the solid surface. In the second step the primary electron with reduced energy excites a valence electron (but with the lower binding energy than in the left panel) and two electrons with energies E_1 and E_2 can leave the solid. It is noted that at each step of scattering the exchange of electrons may occur. An example of a single-step scattering with exchange is shown in Fig. 2.21.

One can see that the primary electron and the valence electron have been interchanged, i.e. the valence electron has final energy E_1 and primary electron has final energy E_2. This process is important when considering the interaction of spin-polarized electrons with a ferromagnetic surface and may lead to an intensity asym-

Fig. 2.21 The final states of the primary electron and the valence electron interchanged

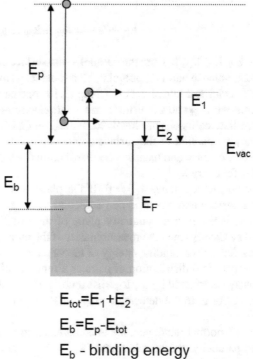

Single-step scattering with exchange

$$E_{tot}=E_1+E_2$$
$$E_b=E_p-E_{tot}$$
$$E_b \text{ - binding energy}$$

metry. This asymmetry is due to an exchange effect and will be considered in more detail later.

As seen from Fig. 2.20, by detecting just energies of two electrons resulting from the interaction of the incident electron with the surface, it is impossible to distinguish two possible scattering paths: (i) a single-step electron-electron scattering (left-side diagram of Fig. 2.20) and (ii) a two—(or multi-) step electron-electron scattering (right-side diagram of Fig. 2.20). However, the momentum conservation (more precisely, parallel-to-the-surface momentum conservation) enables single-step electron collisions to be distinguished from multi-step collisions resulting in the emission of two time correlated electrons, as follows. Since the momentum of the incident electron points into the surface and the total momentum of the electron pair escaping from the surface points outside the surface, an additional scattering "body" must be involved. This additional "body" is the crystal lattice as a whole. Because the periodicity of the crystal is broken in the direction of the surface normal the crystal can supply (or accept) an arbitrary momentum perpendicular to the surface provided the energy of electrons is conserved. At the same time, the periodicity of the crystal along the surface holds. Therefore, parallel-to-the-surface component of the electron momentum is conserved modulo the reciprocal surface lattice vector. It can be written as follows:

$$\mathbf{k}_{0\|} + \mathbf{k}_{1\|} = \mathbf{k}_{2\|} + \mathbf{k}_{3\|} \pm \mathbf{g}_{\|}, \qquad (2.10)$$

where $\mathbf{k}_{0\|}$, $\mathbf{k}_{1\|}$, $\mathbf{k}_{2\|}$, $\mathbf{k}_{3\|}$, are the parallel to the surface components of the momenta of incident, valence and two outgoing electrons, respectively, and $\mathbf{g}\|$ is parallel to the surface reciprocal lattice vector. Using (2.10) one can check whether two detected electrons with $\mathbf{k}_{2\|}$ and $\mathbf{k}_{3\|}$ result from a single electron-electron collision or additional intermediate collisions are involved. Indeed, in (2.10) for the incident electron one can vary its parallel to the surface component of momentum $\mathbf{k}_{0\|}$ by changing the angle of incidence and measure the distribution of electron pairs as a function of $\mathbf{k}_{2\|}$ and $\mathbf{k}_{3\|}$ for every $\mathbf{k}_{0\|}$.

For normal incidence $\mathbf{k}_{0\|} = 0$. If the plane, which contains the normal to the sample surface and which is perpendicular to the scattering plane (containing $\mathbf{k}_{2\|}$ and $\mathbf{k}_{3\|}$) is the mirror symmetry plane of the experimental set-up (including the crystal symmetry) then the measurements of the number of pairs as a function of $\mathbf{k}_{2\|}$ and $\mathbf{k}_{3\|}$ for a given binding energy of the valence electron will provide a symmetric with respect to 0 distribution of pairs as a function of $\mathbf{k}_{2\|} + \mathbf{k}_{3\|}$. This distribution is obviously determined by the $\mathbf{k}_{1\|}$- distribution of the valence electron. The reciprocal lattice vector $\mathbf{g}_{\|}$ in the detectable range of $\mathbf{k}_{2\|}$ and $\mathbf{k}_{3\|}$ must be taken as 0 to satisfy (2.10).

For off-normal incidence, the incident electron momentum has a non-zero component parallel to the sample surface $\mathbf{k}_{0\|} \neq 0$. In that case, the pair distribution as a function of $\mathbf{k}_{2\|} + \mathbf{k}_{3\|}$ is not symmetric with respect to zero point but shifted by $\mathbf{k}_{0\|}$. This shift is observed only under the condition that the two outgoing electrons (with $\mathbf{k}_{2\|}$ and $\mathbf{k}_{3\|}$) are generated in a single electron collision of the incident electron with a valence electron and no additional (intermediate) electron-electron collision (like on the right panel of Fig. 2.20) is involved. In other words, two outgoing electrons "remember" their initial kinematics, if they are generated in a single electron-electron collision.

2.2.1 Experimental Evidence of the Single-Step Scattering Contribution to the (e,2e) Spectrum

To visualize the criterion of a single-step scattering event, (e,2e) spectra are measured for normal and off-normal incidence (from various samples). Let us consider how the required distribution can be extracted from the measured spectra.

As mentioned above the single- and multi-step electron-electron scattering can be separated experimentally by varying incident electron momentum parallel to the sample surface (see (2.10)). Indeed, in the case of a single scattering event, the

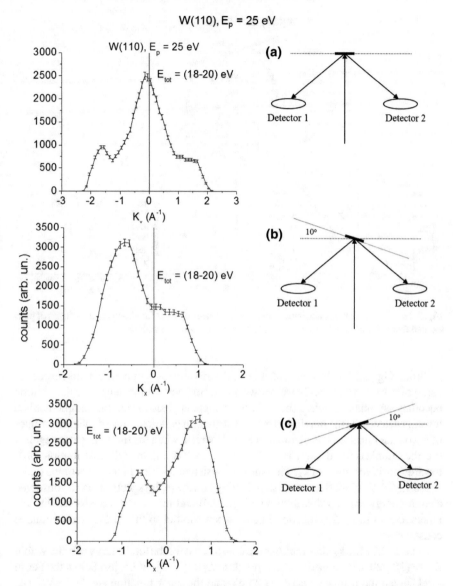

Fig. 2.22 k_x-distributions of electron pairs with total energy (18–20) eV excited from W(110) by 25 eV primary electrons at normal and off-normal incidence

momenta of the incident electron, valence electron, and two outgoing electrons are related by (2.10) and if $k_{0\parallel}$ changes, $k_{2\parallel}$ and $k_{3\parallel}$ also change.

Figure 2.22 presents k_x-distributions of correlated electron pairs excited by 25 eV primary electrons from W(110) for normal and 10° off-normal incidence within the total energy $E_{tot} = (18-20)$ eV.

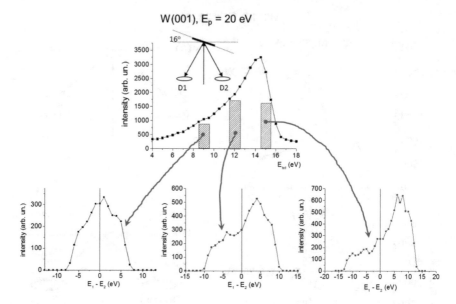

Fig. 2.23 Energy sharing distributions of correlated electron pairs for various total energies of pairs excited from W(001) by 20 eV primary electrons at off-normal incidence

From Fig. 2.22 it is seen that the k_x-distribution is symmetric with respect to zero point for normal incidence. However, for off-normal incidence the distribution becomes asymmetric: when the $10°$ tilting angle is positive (i.e. the sample is tilted towards detector 1) the maximum of the distribution is shifted towards negative values of k_x (on the Fig. 2.22 it is denoted as K_x). Whereas, when the tilting angle is negative (i.e. the sample is tilted towards the detector 2) the maximum of the distribution shifts towards positive values of k_x. This shift is consistent with the kinematics of the (e,2e) scattering. A similar effect of the sample tilting on the (e,2e) spectrum can be observed also in energy sharing distributions (Fig. 2.23) and this is also consistent with the kinematics of the (e,2e) reaction in case when parallel-to-the-surface momentum is conserved.

On Fig. 2.22 the k_x-distributions are constructed for the total energy of pairs within $E_{tot} = (18-20)$ eV range that corresponds to $E_b = (0-2)$ eV i.e. just below the Fermi level (given the primary energy is 25 eV and the work function $e\varphi \approx 5$ eV). The asymmetric distributions for off-normal incidence indicate that electron pairs with this total energy are generated in single-step scattering events. In turn, it means that the energy conservation can be used for the calculation of the binding energy of a valence electron involved in the scattering event.

To answer the question whether electron pairs with a certain total energy are excited in single-step scattering events one needs to analysed k_x-distributions (or energy sharing distributions) for this particular total energy of pairs. An example of such analysis is presented in Fig. 2.23, where it is evident that correlated electron pairs with total energy from 15.5 eV down to 11.5 eV are generated mostly in single-step scattering events. This follows from the fact that the energy sharing distributions at off-normal incidence for these total energies are asymmetric with respect to the zero point. However, for lower total energy (below 9.5 eV in this case) the energy sharing distribution becomes symmetric, which presumably means that correlated electron pairs are generated in multi-step scattering events and they "have forgotten" the initial incident electron momentum. Also for this particular case, electron pairs with total energies within energy band of about $(4 \div 5)$ eV below the onset on the total energy distribution, are generated mostly in single-step scattering events. They correspond to the valence electrons excited from energy levels within $(4 \div 5)$ eV below the Fermi level. Electron pairs with lower total energies are excited mostly in multi-step scattering events. This conclusion holds for most of metal crystal surfaces due to a strong screening (short screening length) of the electron-electron Coulomb interaction. The screening ensures "binary" collision of electrons, i.e. individual electron-electron interaction. In materials with longer screening radius, the "binary" collision conditions hold only for relatively high electron energy, as considered later.

2.2.2   Diffraction of Correlated Electron Pairs on a Crystal Surface

Further evidence of a single-step scattering contribution to the (e,2e) spectrum arises from the observation of the diffraction of correlated electron pairs. We recall first single-electron Low-Energy Electron Diffraction (LEED) which is a phenomenon that occurs when a low-energy electron is elastically scattered from a periodic crystal potential. The diffraction is a result of the wave nature of the electron combined with the periodicity of the scattering potential. The latter implies that the potential can absorb only discrete amounts of momentum determined by a reciprocal lattice vector. In that case, when the wave vector of the incoming electron and the crystal structure are known, the positions of the diffracted beams can be determined from the Bragg diffraction condition. This simple picture of diffraction has to be modified when a composite particle with internal degrees of freedom (and an appropriate wavelength) is scattered from a crystal surface. Then, the change in the internal motion upon the collision has to be taken into account. The description of such a process and experimental examples are presented in (Samarin et al. 2000 Surf Sci). Here we follow this description. The simplest example of this situation is the propagation of an excited, correlated electron pair through a periodic crystal potential out to the vacuum where both electrons can be detected. The internal electron-electron scattering (mediated by the electron-electron interaction) implies that only the total wave vector of the pair is a good quantum number whereas the individual electron's wave vectors are not. When the electron pair is scattered from a periodic potential, the total wave vector of the pair may change by a multiple of the reciprocal lattice vector of the periodic structure. Therefore, it is more appropriate to consider the diffraction of the electron pair rather than the diffraction of the individual electrons of the pair.

It should be emphasized here that this observation is independent of the specific scattering mechanism of the electron pair from the crystal. It relies only on the fact that the electrons may change their individual wave vectors during the collision for reasons other than the elastic scattering from the crystal (here, due to the internal electron-electron correlation) while the total wave vector of the pair is invariant up to a multiple of the reciprocal lattice vector. In fact, it can be shown that when the electron-electron interaction is "switched off" the pair diffraction reduces simply to a diffraction of each of the individual electrons, independently. Thus, it can truly be argued that an observation of the electron pair diffraction is a manifestation of electronic correlation. It is now shown here how the electronic correlation reveals itself in the diffraction of the electron pair and is discussed which information can be extracted from analysis of the diffraction patterns.

The excited electron pair can be generated by an electron or a photon impact (Hermann et al. 1998; Samarin et al. 1998; Iacobucci et al. 1995b). In the electron-pair generation by an impinging low energy electron beam, the incident electrons have energy in the range of 20–100 eV. The two excited electrons escape from the sample and are detected in coincidence in the same hemisphere that contains the incident beam (this scattering geometry is called the back-refection mode). The electronics and low intensity of the primary electron beam ensure that only those electron pairs are collected that are emitted simultaneously after the impact of one single electron. When the electron pair emerges into the vacuum, the (asymptotic) energies and emission angles of the electrons can be resolved (measured).

2.2.2.1 Theoretical Description of the Correlated Electron Pair Diffraction

Consider a low-energy incident electron with momentum $\mathbf{k}_0$ generating an excited electron pair. The two correlated electrons propagate out to the vacuum and arrive at the detectors with wave vectors $\mathbf{k}_1$ and $\mathbf{k}_2$. The Hamiltonian describing the incoming electron beam and the solid surface can be written in the form $H = H_s + H_{ee} + H_{es}$. For simplicity, our discussion does not consider the plasmon and the phonon fields, i.e., the frequencies of the incoming and outgoing electrons are off-resonance with the plasmon frequencies of the metallic surface and additional inelastic processes are neglected. The Hamiltonian of the undisturbed surface is H_s and it describes the electrons in the surface static potential whereas H_{ee} is the interaction within the excited electron pair. The interaction of the incoming electron with the crystal surface is indicated by H_{es}. The spin-dependent transition operator T^S can be derived (within certain approximations specified in Berakdar 1999): $T^S = (1 + (-1)^S P_{12})T$, where S is the total spin of the electron pair (with $S = 1$ ($S = 0$) corresponding to the triplet (singlet) channel) and P_{12} is a permutation operator that exchanges the two excited electrons. The T operator has the form: $T = H_{es}G_0^- + H_{ee}(1 + G_{ee}^- H_{ee})$, where G_{ee}^- and G_0^- are the electron—electron and the free propagators, respectively. Thus, for the numerical evaluation of the transition amplitudes one needs to calculate the matrix elements: $\langle \mathbf{k}_1, \mathbf{k}_2 | T | \mathbf{k}_0, \chi_{\varepsilon(\mathbf{q})} \rangle$. Here $\chi_{\varepsilon(\mathbf{q})}$ is a single-particle eigenstate of H_s with energy ε and crystal momentum $\mathbf{q}$, i.e. a wave function of a bound electron in the valence state.

Since the diffraction of the pairs is considered, it is more appropriate to transform from electron momentum space $\mathbf{k}_1$ and $\mathbf{k}_2$ to the space spanned by $\mathbf{K}^+ \otimes \mathbf{K}^-$, where $\mathbf{K}^+ = \mathbf{k}_1 + \mathbf{k}_2$ is the centre of mass wave vector of the pair. $\mathbf{K}^- = \frac{1}{2}(\mathbf{k}_1 - \mathbf{k}_2)$ is the inter-electronic relative wave vector that describes the internal degree of freedom of the pair (this is only valid under the assumption that the electronic interaction is mainly dictated by the inter-electronic distance).

The periodicity of the scattering potential in the layers parallel to the crystal surface implies Bloch's theorem for the two-particle state. This in turn leads to the conclusion that regardless of the actual functional form of H_{es}, the transition amplitudes $\langle \mathbf{k}_1, \mathbf{k}_2 | T | \mathbf{k}_0, \chi_{\varepsilon(\mathbf{q})} \rangle$ can be expressed as (Berakdar et al. 1998; Berakdar and Das 1997):

$$\langle \mathbf{k}_1, \mathbf{k}_2 | T | \mathbf{k}_0, \chi_{\varepsilon(\mathbf{q})} \rangle = c \sum_{l, \mathbf{g}_{||}} \delta^{(2)} \left(\mathbf{k}_{0||} + \mathbf{q}_{||} + \mathbf{g}_{||} - \mathbf{K}_{||}^+ \right) \times L \left(\mathbf{g}_{||}, l, \mathbf{K}^+, \mathbf{K}^-, \mathbf{q} \right)$$
$$+ \delta^{(2)}(\mathbf{k}_{0||} + \mathbf{q}_{||} - \mathbf{K}_{||}^+) \times L' \qquad (2.11)$$

Here $\mathbf{g}_{||}$ is a surface reciprocal lattice vector. The functions c, L, L' depend on the description of the momentum-space wave function $\chi_{\varepsilon(\mathbf{q})}$ of the bound electron and on the functional form used for the crystal surface potential H_{es}. For the following numerical calculations, a muffin-tin potential for H_{es} is used as described in (Berakdar et al. 1998; Berakdar and Das 1997) and expand G_{ee} to first order in the electron-electron interaction H_{ee}. For the latter coupling, a screened Coulomb potential is used with the screening constant being estimated from the Thomas-Fermi model. Regardless of these limitations for the calculations of L and L' in (2.11), some important conclusions can be drawn from the functional form of (2.11):

(1) Only the centre-of-mass wave vector $\mathbf{K}_{||}^+$ of the pair enters into the Bragg diffraction condition, expressed by the delta function in (2.11). This is equivalent to the diffraction of a quasi-particle located at the pair's centre of mass when $\mathbf{K}_{||}^+$, the parallel component of its wave vector, is changed by $\mathbf{g}_{||}$ during the collision. Note that in LEED studies diffraction occurs when the change in the wave vector $\mathbf{k}_{0||}$ of the incident electron matches $\mathbf{g}_{||}$ (Pendry 1974; Van Hove et al. 1986). The decisive difference to the pair's diffraction is that a fixed $\mathbf{K}^+$ does not imply fixed $\mathbf{k}_1$ and $\mathbf{k}_2$ since a momentum exchange between the two electrons (the internal coordinate $\mathbf{K}^-$ changes then) does not necessarily modify $\mathbf{K}^+$. This may occur, for example, if the two electrons change their individual wave vectors because of mutual repulsion, while their common centre of mass maintains the value of $\mathbf{K}^+$.

(2) While $\mathbf{K}^+$ determines the positions of the diffraction peaks, the functional dependence of L and L' on $\mathbf{K}^-$, which characterizes the strength of the electronic

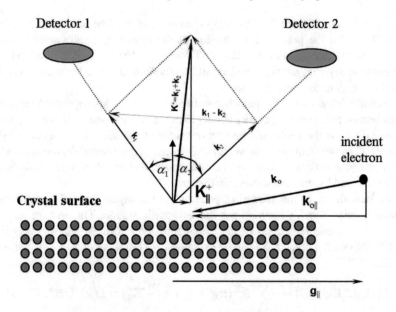

Fig. 2.24 The geometrical arrangement of the experiment for grazing incidence—illustration of the diffraction condition for an electron pair. is the total momentum of the pair, $\mathbf{k}_1$ and $\mathbf{k}_2$ are the momenta of the individual electrons, $\mathbf{g}_\parallel$ is the surface reciprocal lattice vector, $\mathbf{k}_0$ is the momentum of the incident electron with its component $\mathbf{k}_{0\parallel}$ parallel to the surface. Without the participation of $\mathbf{g}_\parallel$ everything would be strongly peaked in the forward direction and no diffracted pair would be detected. (Reprinted from Samarin et al. (2000), Copyright (2000), with permission from Elsevier)

correlation W_{ee} (in the present approximation, W_{ee} depends only on $\left|\mathbf{K}^-\right|$ in reciprocal space), to control the intensity of the individual diffraction peaks. Furthermore, the shape of the individual peaks is influenced by the inter-electronic correlation because the cross-section of the (e,2e) reaction depends strongly on the amount of screening in the Coulomb interaction.

(3) The wave vector $\mathbf{q}\parallel$ of the initially bound Bloch electron is not fixed in the experiment and may vary from 0 to $\mathbf{k}_F$ (Fermi wave vector). This variation results in a broadening of the diffraction pattern even in the case where $\mathbf{K}^+$ and $\mathbf{k}_0$ are experimentally sharply resolved.

(4) Conversely, in the case where $\mathbf{K}^+_\parallel$, $\mathbf{g}_\parallel$ and $\mathbf{k}_{0\parallel}$ are well defined, the position and widths of the diffraction peaks reflect the character of $\mathbf{q}_\parallel$. For example, the maximum width of the diffraction peak is defined by $\mathbf{k}_F$, whereas $\mathbf{K}^+_\perp$, $\mathbf{g}_\perp$, $\mathbf{k}_{0\perp}$ are still not defined.

In the arrangement of Fig. 2.24 the energy conservation law implies a limit for the variation of the wave vector $\mathbf{K}^+_\parallel$, i.e.

$$-\sin\alpha_1\sqrt{2E_{\text{tot}}} \le \left|\mathbf{K}^+_\parallel\right| \le \sin\alpha_2\sqrt{2E_{\text{tot}}}, \tag{2.12}$$

where $E_{tot} = E_1 + E_2$ is the sum energy of the two electrons, the angles α_1 and α_2 determine the positions of the detectors with respect to the surface normal and are shown in Fig. 2.24.

The cut-off condition (2.12) restricts the width of the diffraction peaks, in particular when a diffraction beam occurs at the edges of the wave vector interval given by (2.12).

To outline the role of a pair diffraction the (e,2e) spectra were measured under different experimental conditions: different incident electron energies, various angles of incidence and detection, different lattice constants in the scattering plane (same crystal but different orientation). The measurements were performed for different crystals: W(001), Fe(110), Cu(001). The results are discussed in the following section.

2.2.2.2 W(001), Grazing Incidence

To uncover the contribution from the electron pair diffraction in the (e,2e) spectra the experiments in a grazing-incidence mode were performed where the diffraction condition is fulfilled with $\mathbf{K}_{||}^{+}$ being in the experimentally accessible range (2.12).

For this experiment a W(001) surface was used with an angle of incidence of $2°$ with respect to the surface. The (01) direction of the crystal surface is in the scattering plane with the geometry of the experiment shown in Fig. 2.24 and the incident electron energy was varied in the range of (16–22) eV.

In this situation, the diffraction condition for electron pairs is fulfilled for $\mathbf{K}_{||}^{+}$ being near the middle of the accessible range of $\mathbf{K}_{||}^{+}$. Varying the primary-electron energy (the wave vector $\mathbf{k}_0$ is then varied) one can control the position of the diffraction maximum in the (e,2e) spectrum. Figure 2.25 shows the (e,2e) spectra for different primary-electron energies E_0 but for the same valence electron binding energy $E_b = (-1 \pm 0.5)$ eV relative to the Fermi level. The arrow in each spectrum corresponds to that $\mathbf{K}_{||}^{+}$, which fulfils the diffraction condition (2.11) with $g|| = (-1, 0)$ and assuming $q|| = 0$ ("clean" diffraction condition). In reality, the valence electron, which participates in the collision, may have $q||$ in the range from 0 to $\pm \mathbf{k}_F$. This broadens the diffraction peak by $2\mathbf{k}_F$.

2.2.2.3 Fe(110), Near Normal Incidence

As mentioned above the correlated electron pairs can be detected only when the component of the total wave vector of the pair parallel to the surface belongs to the experimentally accessible range given by (2.12). Varying the angles α_1 and α_2 one can change the accessible range of $\mathbf{K}_{||}^{+}$ and hence change the contribution of the diffracted pairs to the spectrum. Figure 2.26 shows the $\mathbf{K}_{||}^{+}$ distributions of the correlated pair excited from Fe(110), by primary electrons with the energy $E_0 = 50$ eV and with total energy of the pair $E_{tot} = 44 \pm 1$ eV. The angle of incidence is $2°$ with respect to the surface normal. Using the position sensitivity of the detectors, the accessible range of $\mathbf{K}_{||}^{+}$ can be varied by selecting events with larger or smaller angles α_1 and α_2. Vertical

Fig. 2.25 The (e,2e) distributions of correlated electron pairs excited from W(001) at 2° angle of incidence with respect to the surface (geometry is shown in Fig. 2.24). The incident electron energy is E_0 while the total energy of each pair is E_{tot}. Dots are the experimental results. means that the sum momentum vector of each pair points perpendicular to the surface. This does not mean that individual momenta and are equal. (Reprinted from Samarin et al. (2000), Copyright (2000), with permission from Elsevier)

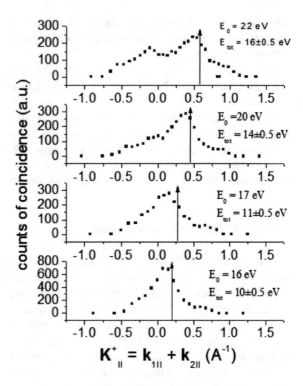

solid and dashed lines show the accessible ranges of $\mathbf{K}_{\parallel}^{+}$ for two sets of measurements respectively: with large angles ($\alpha_1 = 59°$, $\alpha_2 = 53°$) and small ones ($\alpha_1 = 47°$, $\alpha_2 = 41°$). The larger angles correspond to the larger range of $\mathbf{K}_{\parallel}^{+}$ that is accessible for the measurements. The "clean" diffraction conditions (i.e. $\mathbf{q}| = 0$) are shown by arrows, which are outside the accessible range of $\mathbf{K}_{\parallel}^{+}$. Nevertheless, the broadening of the diffraction peaks due to the finite value of the valence electron wave vector (shaded areas in Fig. 2.26) allows to observe the contribution of diffracted correlate pairs. Figure 2.26 shows that narrowing of the accessible range of $\mathbf{K}_{\parallel}^{+}$ (smaller α_1 and α_2) leads to a lower contribution of diffracted pairs.

Figure 2.27 shows another (e,2e) spectrum excited from Fe(110), by primary electrons with the energy $E_0 = 50$ eV and total energy of the pair $E_{tot} = 44 \pm 1$ eV like above. The angle of incidence is 5° with respect to the surface normal, $\alpha_1 = 50°$ and $\alpha_2 = 50°$. The events are integrated over the acceptance angles of the detectors ($\pm 13°$). Comparison of the calculated $\mathbf{K}_{\parallel}^{+}$ distribution (solid curve) with the experimental one shows good agreement. The minimum in the middle of the distribution is mainly due to the fact that the triplet cross-section vanishes for near symmetric geometry and vanishing parallel component of the total momentum of the pair (Berakdar and Das 1997; Berakdar et al. 1998). The different contribution from (0, −1) and (0, 1) diffraction peaks is due to the off-normal angle of incidence.

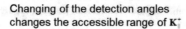

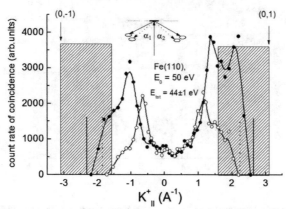

Fig. 2.26 The total parallel momentum spectrum of correlated electron pairs excited from Fe(110) by primary electrons with the energy of 50 eV. The total energy of a pair is $E_{tot} = 44 \pm 1$ eV. Solid circles correspond to the larger angles of the detectors $\alpha_1 = 59°$, $\alpha_2 = 53°$ (shown in the inset). Open circles correspond to the lower angles of detectors $\alpha_1 = 47°$, $\alpha_2 = 41°$. Solid and dashed vertical lines show the limits of corresponding to the two sets of angles. Shaded areas represent the broadening of the diffraction peaks due to the finite value of the valence electron momentum. Vertical arrows denote the "clean" diffraction conditions, i.e. for $q_\| = 0$. (Reprinted from Samarin et al. (2000), Copyright (2000), with permission from Elsevier)

Fig. 2.27 Comparison of the calculated total parallel momentum spectrum (solid line) and the experimental one (dotted line) for Fe(110) excited by primary electrons with 50 eV energy. $E_{tot} = 44 \pm 1$ eV. Shaded areas have the same meaning as in Fig. 2.26. (Reprinted from Samarin et al. (2000), Copyright (2000), with permission from Elsevier)

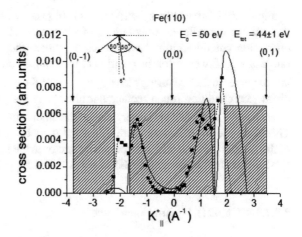

Another way to change the contribution of the diffracted correlated pairs is to change the "clean" diffraction condition by changing the lattice constant in the scattering plane and keeping the accessible range of $\mathbf{K}_\|^+$ constant. The two-fold symmetric (110) surface of iron allows to change the lattice constant in the scattering plane by rotating the sample about the surface normal by 90°.

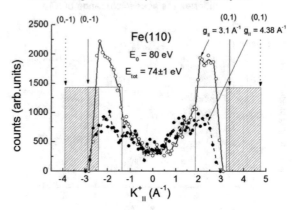

Fig. 2.28 Total parallel momentum spectra of correlated electron pairs excited from Fe(110) by primary electrons with 80 eV energy. $E_{tot} = 74 \pm 1$ eV. Solid circles and open circles correspond to two different azimuthal positions of the sample. $\mathbf{g}_{\parallel} = 3.1$ Å^{-1} corresponds to scattering within the (001) plane of the crystal, $\mathbf{g}_{\parallel} = 4.38$ Å^{-1} to the (10) plane as the scattering plane. Solid and dotted vertical arrows denote the "clean" diffraction conditions for these two positions of the sample. Shaded and open rectangles show the broadening of diffraction maxima due to the finite value of the valence electron momentum. (Reprinted from Samarin et al. (2000), Copyright (2000), with permission from Elsevier)

Figure 2.28 shows two spectra for Fe(110) taken with $E_0 = 80$ eV and $E_{tot} = 74 \pm 1$ eV but for different orientations of the sample. The arrows show the "clean" diffraction conditions for these two cases, where they are beyond the accessible range of $\mathbf{K}_{\parallel}^+$. However, for the position of the sample corresponding to the reciprocal lattice vector in the scattering plane $\mathbf{g}_{\parallel} = 3.1$ Å^{-1} the overlap between the broadened diffraction condition (open rectangular area) with the accessible range of $\mathbf{K}_{\parallel}^+$ (that spans from -3 to 3 Å^{-1}) is larger than for the other sample position ($\mathbf{g}_{\parallel} = 4.38$ Å^{-1}). Therefore, the contributions from diffracted pairs becomes substantial for the case of the sample position with the reciprocal vector in the scattering plane $\mathbf{g}_{\parallel} = 3.1$ Å^{-1}: two maxima in the spectrum at $\mathbf{K}_{\parallel}^+ = -2$ Å^{-1} and $\mathbf{K}_{\parallel}^+ = 2.5$ Å^{-1} arise.

2.2.2.4   Cu(001), Normal Incidence

Another example of diffraction of correlated electron pairs is the (e,2e) spectrum measured on Cu(001) surface that was excited by 85 eV primary electrons at normal incidence. Figure 2.29 shows momentum sharing distribution at total energy $E_{tot} = 79 \pm 1$ eV. The theoretical curve (solid line) was calculated for infinite energy- and angle-resolution of the detectors. For clarity the theoretical $(0, -1)$ and $(0, 1)$ diffracted peaks are scaled down by factor of 2. The shaded areas show again the broadening of the diffraction peaks due to the variation of the wave vector of the

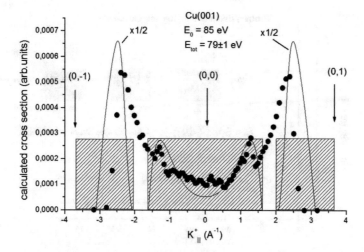

Fig. 2.29 Momentum-sharing spectrum for Cu(001) excited by primary electrons with 85 eV energy. $E_{tot} = 79 \pm 1$ eV. Solid line—calculated spectrum, dots—experiment. Diffraction peaks in the calculated spectrum are scaled down by 2. Shaded areas show the range of broadening of diffraction maxima due to the finite value of the valence electron momentum. (Reprinted from Samarin et al. (2000), Copyright (2000), with permission from Elsevier)

valence electron of the target. The calculated spectrum reproduces fairly well the general shape of the experimental one.

2.3 Surface Sensitivity of the Low-Energy (e,2e) Scattering in a Back-Reflection Geometry

One of the important characteristics of a surface science technique is its surface sensitivity or probing depth. Most of these techniques are based on the interaction of particles (electrons, photons, ions, atoms) with the surface and then observation of the response of the surface.

Usually, the response is observed in the form of emission of secondary particles (electrons, photons, ions). It is interesting to compare the surface sensitivity of (e,2e) spectroscopy with Electron Energy Loss Spectroscopy (EELS) and photoelectron spectroscopy (PES) since they are all based on the interaction of electrons with surfaces. In PES, a photon impact on the surface induces an electron emission. The penetration depth of the photon, when estimated roughly by the wavelength of the photon, is much larger than the mean free path (MFP) of the excited electrons and, in that way, the MFP of the photoelectrons characterizes the probing depth of PES. In EELS two electrons are involved in the process of excitation and, for simplicity, it is assumed that the incident and scattered electrons have the same MFPs, say λ. Then the probing depth of the EELS is approximated as $\lambda/2$ while noting the energy

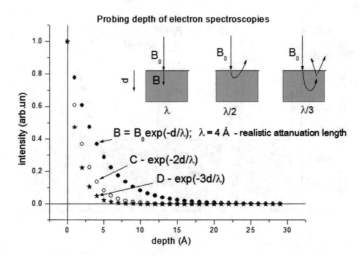

Fig. 2.30 Probing depth of the (e,2e) spectroscopy is estimated as $\lambda/3$, where λ is the mean-free path of an electron

dependence of λ for various materials (Hüfner 1996). Then for the case of (e,2e) spectroscopy, *two* low-energy electrons have to escape from the surface and hence three electrons are involved in the scattering event. Therefore, the surface sensitivity is increased as compared to SPEELS. Indeed, if the mean-free path of the electrons involved in the (e,2e) reaction is on average λ, then the probing depth of the (e,2e) spectroscopy is estimated to be $\lambda/3$ (see Fig. 2.30).

2.3.1   Contribution of Surface States on W(100) to the (e,2e) Spectrum Measured with Low-Energy Primary Electrons

For a single crystal sample, it was shown (Iacobucci et al. 1995a; Samarin et al. 1997; Artamonov et al. 1997) that the experimental low-energy (e,2e) data are consistent with a two-step model, assuming that an elastic diffraction process leads to the back reflection of the primary electron as a first step, and is followed by an electron–electron collision in the second step. With that model one can consider the incident electrons first elastically backscattered at the lattice and then generating correlated electron pairs which leave the sample in a backward direction. The energy and momentum of the incident (diffracted) electron just before the electron–electron collision are known. So, the measurement of energy and momenta of the two coincident final electrons allows analysis of the scattering dynamics of the process.

For low energy primary electrons, the theoretical treatment of this process requires the initial and final state wave functions to be described by LEED (Low Energy

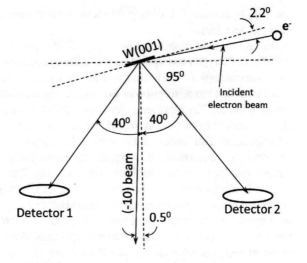

Fig. 2.31 Sketch of the experimental setup. (Reprinted from Samarin et al. (1998), Copyright (1998), with permission from Elsevier)

Electron Diffraction) states (Gollisch et al. 1997), which leads to a more complicated relation between the differential (e,2e) cross section and the momentum density distribution of the valence electrons than in the case of transmission geometry.

The goal of this paragraph is the evaluation of the contribution of emitted electron pairs originating from the surface states to the total emission of correlated pairs. One can extract $q_{\|}$-distribution functions for W(001) surface states from our data if the (e,2e) cross section to be proportional to the target electron momentum distribution. By comparing the $q_{\|}$-distributions with those obtained from photoemission measurements, an 'order of magnitude' estimate is obtained of the deviation of the assumed proportionality between the (e,2e) cross section and the initial momentum distribution for the case of the geometry and energy range of the (e,2e) experiment. To enhance the surface sensitivity the grazing geometry of scattering was chosen. A sketch of the setup is shown in Fig. 2.31 with two 75 mm MCP-based position sensitive electron detectors oriented in a plane that contains both detector axes and the surface normal of the sample. Each detector is 140 mm from the sample with an angular acceptance in the scattering plane of $\pm15°$ and the two detector axes are at an angle of 80° to each other. A parallel electron beam of about 1 mm diameter impinges on the sample surface such that it includes an angle of 2.2° with the surface and angle of 95° with the bisector between both detectors, the electron gun being in the plane of the detectors.

Pairs of electrons, generated in the sample by the primary electron beam that leaves the sample in a backward direction, are detected in coincidence by the two detectors. The primary electron energy is 20 eV and the energy resolution is 0.4 – 0.7 eV. Clean surface conditions of the W(001) sample during the measurement are provided by standard cleaning and analysis procedures. Orientation of the sample is achieved by a rotatable holder where each particular setting is controlled by observing the LEED-patterns of elastically diffracted primary electrons in the position sensitive electron detectors.

To visualize the manifestation of the W(001) surface states in the (e,2e) spectra the effect of oxygen adsorption on the (e,2e) spectra was studied. Using photoemission spectroscopy (Feuerbacher and Fitton 1972; Feuerbacher and Willis 1976) it was found that the adsorption of O_2 suppresses the observation of W(001) surface states. Figure 2.32a represents the summed (total) energy distribution of electron pairs measured from a clean W(001) surface (solid circles) in comparison with the corresponding distribution obtained from the oxygen covered surface (open circles). The spectra are normalized to equal acquisition time, while the incident electron current was kept constant for both measurements. The difference between the two spectra is represented by solid circles in Fig. 2.32b, where the summed energy of correlated electron pairs is converted into binding energy of the initial valence state with respect to the Fermi level using the primary electron energy and the W(001) work function (4.6 eV). The difference spectrum shows two pronounced maxima at the binding energies $(E_1)_{e2e} = -0.4$ eV and $(E_2)_{e2e} = -4.4$ eV. These energies coincide fairly well with the energies of two surface states S_1: $(E_1)_{PE} = -0.3$ eV and S_2: $(E_2)_{PE} = -4.2$ eV found in the photoemission spectrum of W(001) (Weng et al. 1978) that is represented by open squares in Fig. 2.32b. Although the relative intensities of the two maxima are different for the photoemission and (e,2e) spectra, one can take the coinciding energy peak positions as identifying signatures for the W(001) surface states, contributing to the (e,2e) cross section. It should be noted that the maxima in the (e,2e) difference of Fig. 2.32b are considerably broader than those in the photoemission spectrum, which can be attributed to a higher energy and angular resolution in the photoemission measurement than in the (e,2e) spectroscopy. The difference between the summed-energy spectra for the clean and oxygen covered W(001) surface indicates that the oxygen layer on the W(001) surface strongly suppresses the generation of correlated electron pairs from surface states. In general, these results demonstrate the high surface sensitivity of low energy back-reflection (e,2e) spectroscopy.

In Fig. 2.32c the energy sharing distribution of electron pairs with summed energies $E_s = 14.5 \pm 1$ eV, corresponding to the initial valence state energy related to the surface state S_1, is presented. The figure shows the distributions obtained from the clean W(001) surface (full circles) and from the 0.1 L O_2-covered surface (open circles). The spectrum measured on the clean surface shows a pronounced peak at $E_1 - E_2 = 3$ eV on top of a nearly symmetric distribution, while this peak is almost invisible in the distribution from the oxygen covered surface. As shown in Fig. 2.32d, the difference between both spectra illustrates the additional contribution of the clean surface. In accordance with the analysis of the summed energy distributions (Fig. 2.32a/b), one can identify the electron pairs represented by the distribution in Fig. 2.32d as the contribution of W(001) surface states to the measured (e,2e) cross section (from the clean surface). Figure 2.32d shows that electron pairs originating from S_1-states share their summed energy of $E_s = 14.5 \pm 1$ eV most likely in the fractions of $E_1 = 9 \pm 05$ eV and $E_2 = 5.5 \pm 0.5$ eV. According to the two-step model for the low energy back-reflection (e,2e) process, one can con-

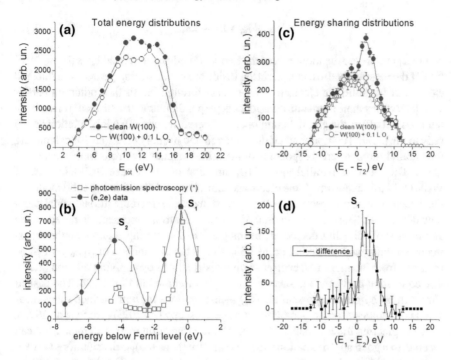

Fig. 2.32 Effect of the oxygen adsorption on the energy distribution of correlated electron pairs. **a** Summed energy distribution of electron pairs from the clean (full circles) and oxygen covered (open circles) W(001) surface. **b** The difference between the energy summed distributions shown in (**a**) (full circles) in comparison with the energy distribution of two W(001) surface states (S1, S2) measured by photoemission (Shin et al. 1997). The abscissa represents the binding energy with respect to the Fermi level. **c** Energy sharing distribution of electron pairs related to the surface state S1 from the clean (full circles) and oxygen covered (open circles) W(001) surface. **d** Difference between the two distributions shown in (**c**). (Reprinted from Samarin et al. (1998), Copyright (1998), with permission from Elsevier)

sider the correlated electron pairs to be generated by primary electrons that have been elastically diffracted at the lattice in the first step. In the present case, the diffraction of 20 eV primary electrons at the W(001) crystal surface leads to an elastic '(−10) diffracted electron beam' in a direction close to the bisector between both detectors (see Fig. 2.31). By turning the W-sample towards detector 1 or detector 2, we can observe this diffraction maximum as a spot on the corresponding detector, thus enabling verification of the (−10)—beam position for the particular sample setting and primary energy during the measurement as shown in Fig. 2.31. According to photoemission results (Shin et al. 1997), the major contribution of correlated electron pairs observed in this geometry can be considered to be generated by the (−10) diffracted electron beam. Based on this assumption the initial momentum **q** of the valence electron inside the solid is given by the momentum conservation condition:

$$q = k_1 + k_2 - k_0, \tag{2.13}$$

where k_1 and k_2 are the momenta of the two scattered electrons and k_0 is the momentum of the elastically diffracted electron inside the solid. During penetration into and exit out of the solid the electrons have to pass through the surface potential barrier which modifies their momenta components perpendicular to the surface (refraction), but does not affect the parallel components. Therefore, one can apply relation (2.13) only to the parallel components of the scattered and primary electrons for calculating the parallel momentum components of the initial valence electrons in the scattering plane. This plane is parallel to the [10] direction of the surface Brillouin zone of W(001). In this experiment, the electrons' emission angles within the solid angles of the detectors were not resolved, but integrated in order to increase the statistical accuracy of the data. Thus, determination of the final electronic momenta is based on the assumption of point-like detectors (the integration of scattering angles over the experimental angular acceptance). Figure 2.33a shows the distribution of $(q_\parallel)_{e2e}$ for S_1 as obtained from the measured energy sharing distributions by applying relation (2.13). The equivalent curve for the surface state S_2 is presented in Fig. 2.33b. The dashed curves in both figures represent the $(q_\parallel)_{PE}$-distributions for both surface states measured with angle-resolved photoemission spectroscopy (ARPES) (Weng et al. 1978). When neglecting the k-dependence of the dipole transition matrix elements, we can assume, to a good approximation (Straub et al. 1997), the $(q_\parallel)_{PE}$-distributions as measured by ARPES to be proportional to the parallel momentum distribution of the two surface states S_1 and S_2. In both figures, especially in Fig. 2.33b, there is a reasonable agreement of the relative shapes between the $(q_\parallel)_{e2e}$ and $(q_\parallel)_{PE}$ distributions, while in Fig. 2.33a the (e,2e) distribution is shifted by about 0.2Å^{-1} towards lower parallel momenta and in Fig. 2.33b the half-width of the peak at $(q_\parallel) = 0.2 \text{Å}^{-1}$ is about twice as large in the (e,2e) curve ($\Delta q_\parallel \approx 0.25 \text{ Å}^{-1}$) as in the distribution obtained from ARPES. These differences are partly due to the fairly low momentum resolutions in the $(q_\parallel)_{e2e}$ distributions. Considering the acceptance angle of the electron detectors and the experimental energy resolution, it is estimated that the resulting momentum uncertainty in the calculated $(q_\parallel)_{e2e}$ distributions is approximately 0.25Å^{-1}.

In spite of the described discrepancies, there is a general similarity of the parallel momentum distribution of the W(001) surface states obtained from ARPES measurements with the $(q_\parallel)_{e2e}$ distributions of these states extracted from low energy back-reflection (e,2e) data. This finding suggests that the momentum distribution of the initial valence electrons determines the main features of the differential cross section for low energy back-reflection (e,2e) processes.

Fig. 2.33 $q_{\|}$-distributions of the surface states S_1 (**a**) and S_2 (**b**) deduced from the (e,2e) measurement (full circles) in comparison with those obtained from photoemission measurements (open circles). (Reprinted from Samarin et al. (1998), Copyright (1998), with permission from Elsevier)

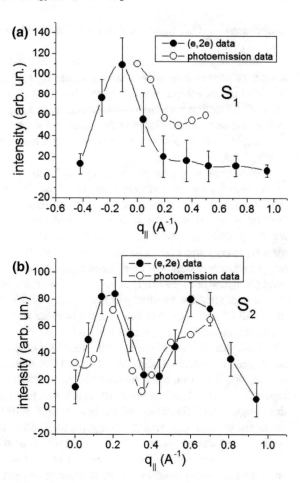

2.3.2 Observation of Oxygen Adsorption on W(110) by Low-Energy (e,2e) Spectroscopy

Adsorption of oxygen on the (001) surface of tungsten is well studied by photoelectron spectroscopy, LEED, Auger spectroscopy, EELS and other techniques (Bradshow et al. 1974; Avery 1981; Luscher and Propst 1977; Mullins and Overbury 1989; Weng et al. 1978). It was found that oxygen molecules undergo dissociative adsorption on tungsten. At small coverages the oxygen atoms are well ordered on the W(001) surface and the oxygen layer presents a variety of structures which are dependent on the temperature of the sample during adsorption, the temperature treatment after adsorption and the amount of oxygen atoms (Brashow et al. 1974). It was found also that oxygen adsorption changes the electronic structure of the surface which is manifested in photoelectron spectra (PES), some of the maxima in PES disappeared after adsorption and, on the other hand, new maxima appear (Brashow et al. 1974).

In order to study the emission of correlated electron pairs from adsorbed oxygen a layer of oxygen was adsorbed on a clean W(001) surface at exposure of 10 L, followed by the sample heating up to 1400 °C. This procedure lead to the appearance of diffraction patterns corresponding to the superlattice of oxygen with the lattice constant twice as large as the lattice constant of W(001). This exposure and the sample heating provide the ordered oxygen layer on W(001) with the structure $(1 \times 2, 2 \times 1)$.

Test measurements were performed with the different positions (orientations) of the sample, i.e. with different angles of incidence, and it was found that the maximum count rate of coincidences is observed at the incident angle of 22.5°. This "optimum" position of the sample was used for further measurements as well as a grazing incidence (with 2.2° of incidence) and normal incidence. The incident electron energy was chosen to be equal to 36 eV because the binding energy of the oxygen adsorbate state was found to be equal to about 6 eV with respect to the Fermi level (Brashow et al. 1974), i.e. about 11 eV with respect to the vacuum level.

Consequently the incident electron energy of 36 eV is well above the energy threshold of the (e,2e) reaction in the adsorbate state. The energy diagram of the correlated electron pair generation from the oxygen adsorbate state is presented in Fig. 2.34. An incident electron with energy E_0 may excite a bound electron from the oxygen 2p—derived adsorbate state and two electrons with energies E_1 and E_2 may escape the surface and be detected in coincidence by two electron detectors. Figure 2.35 shows total energy distributions (left panel) and energy sharing distributions (right panel) for clean and oxygen covered W(001) surfaces. The spectra are measured at three geometrical arrangements that are depicted next to spectra and with primary electron energy of 36 eV. For all geometries, an additional maximum at the total energy of 24.5 ± 1 eV appears in the total energy distributions and its amplitude gradually increases with increasing oxygen coverage (Fig. 2.35a). The energy width of this maximum reflects the energy broadening of the 2p state of atomic oxygen due to coupling to the substrate. This result is substantiated by the fact that the position and the width of the maximum associated with oxygen adsorption is virtually not changing for different scattering geometry (cf. Fig. 2.35a–c). On the other hand the total energy spectrum of clean W(001) is very dependent on the geometrical arrangement of the scattering experiment because electron pairs emitted from tungsten are subject to inelastic scattering processes that are very sensitive to the scattering geometry. Each pair of spectra, clean W(001) and W(001)+oxygen, are normalized by the acquisition time the incident current being constant.

The total energy distributions do not resolve energies E_1 and E_2 of individual electrons and contain only integral information on the electron emission process. More insight into the emission process is obtained by analysis of the energy-sharing distributions of escaping electrons at a given total energy of $E_{tot} = 24.5 \pm 1$ eV which corresponds to electron excitation from the adsorbate state. Therefore, the excitation mechanism from these localized electronic states may be different from the one from delocalized valence state of tungsten, for example. Thus, to obtain

Energy diagram of the oxygen on W(001) adsorbate system

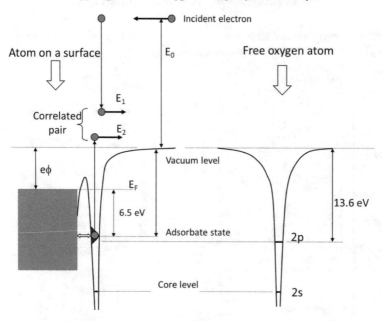

Fig. 2.34 Energy diagram of the correlated electron pair generation from oxygen $2p$-derived adsorbate state. E_0—incident electron energy, E_1 and E_2—energies of electrons of the pair, $e\phi$—workfunction of the surface

some insight into the pair emission from localized states, it is natural to focus here on the differences of spectra for an oxygen-covered tungsten surface and a clean tungsten surface: [W(001)+oxygen] − [W(001)].This difference spectrum should provide clear information on the electron emission from the adsorbate states such as presented in Fig. 2.36.

A theoretical description of the electron pair emission from an adsorbate system should account for the modification of the surface (substrate) and the adsorbed atoms due to their interaction (Samarin et al. 1999). The electron-pair emission probability from the "adsorbate system" was analyzed by J. Berakdar (in Samarin et al. 1999) and is given by:

$$\frac{d\sigma(\mathbf{k}_1, \mathbf{k}_2)}{d^3\mathbf{k}_1 d^3\mathbf{k}_2} = \frac{(2\pi)^4}{k_0} \sum_{\alpha\beta} \int d^3\mathbf{k}_j \rho_{dos}(\varepsilon(\mathbf{k}_j)) F(\varepsilon(\mathbf{k}_j), T) |\Im|^2 \delta(E_f - E_i), \qquad (2.14)$$

where $\mathbf{k}_1$ and $\mathbf{k}_2$ are the wave vectors of the vacuum electrons, $\rho_{dos}(\varepsilon(\mathbf{k}_j))$ is the density of states, k_0 is the projectile's incident momentum, E_i (E_f) is the total energy in the initial (final) state and F is the fermi distribution at the temperature T. Equation (2.14) sums over the final states degeneracy β, and over non-observed initial states α.

W(001) + oxygen, E_p = 36 eV.

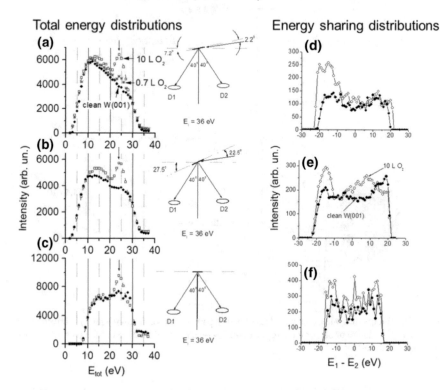

Fig. 2.35 Total energy distributions (left column) of correlated electron pairs excited by 36 eV primary electrons from clean W(001) (solid circles) and oxygen covered W(001) (open circles) measured at various geometries shown next to the distributions. Right column—corresponding energy sharing distributions at $E_{tot} = 24.5 \pm 1$ eV

The transition amplitude is, to first order, the sum of two amplitudes: $\Im = T^{sub} + T^{ad}$, where T^{sub} and T^{ad} describe the pair emission from the substrate and from (2×1) adsorbed oxygen layer.

The amplitude T^{sub} has the form (Bereakdar et al. 1998):

$$T^{sub} = \delta^{(2)}(\mathbf{K}_{0||} - \mathbf{K}_{||}^+)\Lambda' + C \sum_{l,\mathbf{g}_{||}^{sub}} \delta^{(2)}\big[\mathbf{g}_{||}^{sub} - (\mathbf{K}_{||}^+ - \mathbf{K}_{0,||})\big]\Lambda, \qquad (2.15)$$

where functions C, Λ, and Λ' depend on the dynamical description of the processes, l enumerates atomic layers of the substrate, $\mathbf{K}_0 = \mathbf{k} + \mathbf{k}_0$ is the initial wave vector of the pair, $\mathbf{K}^+ = \mathbf{k}_1 + \mathbf{k}_2$ is the final wave vector of the pair, $\mathbf{g}_{||}^{sub}$ is parallel to the surface reciprocal lattice vector of the substrate. The transition amplitude of the adsorbate T^{ad} consists of two parts:

Fig. 2.36 Contribution of
the ordered oxygen
adsorption to the spectrum of
the correlated electron pairs.
Comparison of experiment
and the theory

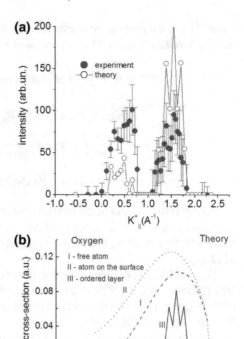

$$T^{\mathrm{ad}} = T^{\mathrm{ee}} + T^{\mathrm{core}}, \tag{2.16}$$

where T^{ee} describes electron-electron scattering, and T^{core} describes electron scattering by ion cores. Since the lattice constant of the oxygen superlattice is larger than the extent of the active atomic orbitals X a tight-binding description for the electronic wave function of the oxygen monolayer can be used:

$$\Phi_{k_{\|}}(\mathbf{r}_2) = N \sum_j \exp(i\mathbf{r}'_{\|} \cdot \mathbf{R}_{\|,j}) \phi(\mathbf{r}_{2\|} - \mathbf{R}_{j\|}, r_{2z} - d). \tag{2.17}$$

Here $\phi(\mathbf{r}_{2\|} - \mathbf{R}_{j\|}, r_{2z} - d)$ is an atomic orbital localized at the site $(\mathbf{R}_{j\|}, d)$, where d is the distance between the adsorbed layer and the surface layer of substrate (in this case $d = 1.25$ Å), $\mathbf{k}'_{\|}$ is a two-dimensional Bloch wave vector and $\mathbf{r}_2$ is the position vector of the bound electron. Following the emission process, we assume the superlattice is neutralized by the substrate on a time scale much faster than the ejection process. In this case the following assumption for the two-electron state is reasonable:

$$|\Psi\rangle = (\mathbf{1} + G_{ee}V_{ee})|\mathbf{k}_1, \mathbf{k}_2\rangle, \tag{2.18}$$

where G_{ee} is the propagator within the potential V_{ee}. Under the above assumptions the amplitudes T^{ee} and T^{core} can be written as:

$$T^{ee} = N_c N \sum_j \int d^3\mathbf{r}_1 d^3\mathbf{r}_2 \exp(i\mathbf{q} \cdot \mathbf{r}_1 - i\mathbf{k}_2 \cdot \mathbf{r}_2)$$

$$\bar{\Psi}(\mathbf{r}_1, \mathbf{r}_2) V_{ee} \exp(i\mathbf{k}'_{||} \cdot \mathbf{R}_{j||}) \phi(\mathbf{r}_{2||} - \mathbf{R}_{j||}, r_{2z} - d) \tag{2.19}$$

$$T^{core} = -N_c \sum_{kl} \int d^3\mathbf{r}_1 d^3\mathbf{r}_2 \frac{Z_t}{|\mathbf{r}_{1||} - \mathbf{R}_{l||}| + |r_{1z} - d|} e^{i\mathbf{q}\cdot\mathbf{r}_1 - i\mathbf{k}_2\cdot\mathbf{r}}$$

$$\bar{\Psi}(\mathbf{r}_1, \mathbf{r}_2) \exp(i\mathbf{k}'_{||} \cdot \mathbf{R}_{k||}) \phi(\mathbf{r}_{2||} - \mathbf{R}_{j||}, \vec{r}_{2z} - d) \tag{2.20}$$

Here N is the number of the oxygen sites, N_c is a normalization constant, Z_t is the core charge, $\mathbf{q} = \mathbf{k}_0 - \mathbf{k}_1$ is the momentum transfer vector, $\mathbf{r}_1$ is the position vector of the incident electron, and $\bar{\Psi}$ is a function that can be deduced from the solution of (2.18).

The multi-center *hopping integrals* involved in (2.20) are quite difficult to calculate. However, as φ is well localized on the scale of the superlattice constant, the main contribution to (2.20) originates from the *on-site* emission, i.e. for $k = l$. Furthermore, it can be assumed that the electron-pair current is the incoherent sum of the currents from the substrate and the adsorbate. Then the cross section for the pair emission becomes proportional to $|T^{sub}|^2 + |T^{ad}|^2$, where $|T^{ad}|^2$ can be obtained from (2.16, 2.19, 2.20):

$$|T^{ad}|^2 = \sum_{\mathbf{g}^{ad}_{||}} \delta^{(2)}[\mathbf{g}^{ad}_{||} - (\mathbf{K}^+_{||} - \mathbf{K}'_{0||})] \cdot |\overline{T^{ee}} + \overline{T^{core}}|^2 \tag{2.21}$$

where $\mathbf{K}'_{0||} = \mathbf{k}_{0||} + \mathbf{k}'_{||}$ and

$$\overline{T^{ee}} = N_c \exp[i(q_z - k_{b,z}) \cdot d] \int d^3\boldsymbol{\rho}_1 d^3\boldsymbol{\rho}_2 \exp(i\mathbf{q} \cdot \boldsymbol{\rho}_1 - i\mathbf{k}_2 \cdot \boldsymbol{\rho}_2) \bar{\Psi}(\boldsymbol{\rho}_1 - \boldsymbol{\rho}_2) V_{ee} \varphi(\boldsymbol{\rho}_2) \tag{2.22}$$

$$\overline{T^{core}} = -N_c \exp[i(q_z - k_{b,z}) \cdot d] \int d^3\boldsymbol{\rho}_1 d^3\boldsymbol{\rho}_2 \frac{Z_t}{\rho_1} \exp(i\mathbf{q} \cdot \boldsymbol{\rho}_1 - i\mathbf{k}_2 \cdot \boldsymbol{\rho}_2) \bar{\Psi}(\boldsymbol{\rho}_1 - \boldsymbol{\rho}_2) \varphi(\boldsymbol{\rho}_2) \tag{2.23}$$

The transitional factor $\exp[i(q_z - k_{b,z}) \cdot d]$ is irrelevant for the cross section. From (2.15 and 2.21), it is clear that there are two types of diffraction: (1) diffraction of the pair from the W(001) lattice with the maximum width of the diffraction patterns given by k_F, where k_F is the Fermi wave vector of the substrate state, and (2) diffraction from the super-lattice of adsorbed atoms with the shape of these diffraction patterns being determined by $|\overline{T^{ee}} + \overline{T^{core}}|^2$, as evident from (2.20). The interference between the electron-pair currents from the substrate and the super-lattice of adsorbed atoms are described by the term $(T^{ad})^* T^{sub} + T^{ad}(T^{sub})^*$.

The pairs diffraction features are most visible in the energy sharing distributions, where the number of correlated electron pairs is plotted as a function of the energy difference $E_1 - E_2$.

As evident from (2.14 and 2.20), the effect of diffraction is most transparent when the spectra are analysed as a function of $K_\parallel^+$. In the experiment the electron-pair current is measured as a function of the component of $K_\parallel^+$ in the plane of the detectors, K_x^+, that is given by:

$$K_x^+ = 0.363\left(\sqrt{E_{\text{tot}} + \Delta}\cos\alpha_1 - \sqrt{E_{\text{tot}} - \Delta}\cos\alpha_2\right),$$

where α_1 and α_2 are the emission angles of the electrons with respect to the sample surface and $\Delta = E_1 - E_2$.

The K_x^+ distributions of the difference spectra, [W(001)+oxygen] – [W(001)], is shown in Fig. 2.36 (solid circles on top panel). Two pronounced maxima are observed at $K_x^+ = 0.5$ Å^1 and $K_x^+ = 1.5$ Å^{-1}. Also depicted in Fig. 2.36 are the theoretical results, as yielded by (2.21). The final state is given by (2.18), i.e. the electron-electron interaction is taken to all orders in the final state. For the calculation of the integrals (2.22, 2.24) a single zeta $2p$ state was employed for the wave function φ. The cross section is then obtained from (2.14 and 2.21) after averaging over initial state magnetic sublevels and summing over the degrees of freedom of the spins of the emitted pair. The positions of the maxima are reproduced by theory. However, the theory predicts a different ratio of the magnitude of the peaks. The reason for this shortcoming is comprehensible if one considers the pair emission from a free oxygen atom and from one atom placed on the tungsten surface. The results for both cases are shown in Fig. 2.36b. As clear from Fig. 2.36 the cross section for the electron pair emission from a free or tungsten adsorbed oxygen atom is small around the region of the smaller peak in the spectrum of pair emission from one monolayer. Thus, one can conclude that the origin of the asymmetric heights of the theoretical peaks in Fig. 2.36a can be traced back to the behaviour of the atomic ionization cross section, as depicted in Fig. 2.36b.

Further effects to be included in the theory are the distortion and dispersion of the $2p$ symmetry of the initially bound state and the inclusion of the next order hopping integrals.

As mentioned above, there are two contributions to the diffraction patterns from adsorbate system: diffraction of the pair from the W(001) lattice and diffraction from the super-lattice of adsorbed atoms. Since the lattice constant of the superlattice is twice as large as the lattice constant of the substrate, one can expect to see diffraction patterns from the superlattice at smaller K_x^+. Figure 2.37 presents the K_x^+ distributions of correlated electron pairs from clean W(100) (Fig. 2.37b) and from ordered adsorbate layer of oxygen (Fig. 2.37a). The vertical arrows depict the position of diffracted beams according to (2.15 and 2.21). Two diffracted patterns are clearly observed on the K_x^+ distribution from oxygen layer (Fig. 2.37a) at $K_x^+ = 0.5$ Å^1 and $K_x^+ = 1.5$ Å^{-1}, whereas the second diffraction pattern from the clean W(001) surface ((00) diffraction beam) is beyond the accessible range of K_x^+. Therefore, only the rising part of this maximum is observed at $K_x^+ = (1.5 \div 2)$ Å^{-1}.

Fig. 2.37 Diffraction patterns from adsorbed oxygen atoms (**a**) and from clean W(001) crystal (**b**)

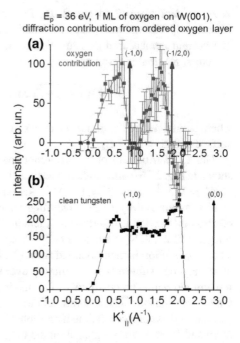

2.4 Plasmon-Assisted (e,2e) Scattering from a Surface

As already mentioned, the secondary electrons generated from a surface by impact of an incident electron, are divided, by convention, in three groups: (i) elastically scattered electrons, i.e. electrons that did not lose energy while scattering from the surface; (ii) inelastically scattered electrons, i.e. electrons that lost part of their energy while scattering from the surface, and (iii) true secondary electrons that are generated in a multi-step scattering process. However, there might be a link between electrons of the second and the third groups. Indeed, one can imagine, for example, a scenario, when the incident electron with the energy E_0 loses part of its energy ΔE for a plasmon excitation. Then the plasmon may decay transferring all its energy and momentum to a valence electron, which may escape from the surface. This process can be directly observed using two-electron coincidence spectroscopy and detecting two electrons with energies E_1 and E_2. One of the electrons would be the incident electron that lost part of its energy $E_1 = E_0 - \Delta E$ for the plasmon excitation, and the second electron will be the ejected one with energy $E_2 = \Delta E - e\varphi$, where $e\varphi$ is the binding energy of the valence electron relative to the vacuum level. The energy of the second electron will be the same independent on the incident electron energy. The energy balance between the two electrons would indicate the mechanism of the scattering reaction.

2.4.1   Secondary Electron Emission from Dielectrics

One of the good candidates to display such a mechanism of the (e,2e) reaction is LiF film. Indeed, in the secondary emission spectrum of LiF an emission feature was observed at the electron energy of 7 eV (Henke et al. 1979; Golek and Bauer 1996; Gusarov and Murashov 1994; Samarin et al. 2004a). The position of the maximum is independent on the primary electron energy and the maximum was suggested to be due to the decay of plasmons excited by the incident electrons (Henke et al. 1979; Golek and Bauer 1996; Gusarov and Murashov 1994). Due to the plasmon de-excitation via electron emission, a peak in the energy distribution curves of secondary electrons appears at the energy E_e. However, the energy and wave-vector conservation rules have to be satisfied for this reaction. The energy conservation requires that

$$E_e = \hbar\omega_p - E_g - lE_v - \chi, \qquad (2.24)$$

where E_g is the energy gap, ΔE_v is the half-width of the valence band and χ is the electron affinity. In the case of a single crystal the wave vector conservation rule implies that the wave vector of the ejected particles has to be equal to the wave vector of the plasmon.

Parameters characterization of a dielectric film (LiF, for example): $\hbar\omega_p$, E_g, ΔE_v, and χ may vary from sample to sample depending on the film preparation procedure, thickness, and type of substrate. Therefore, the energy position of the emission feature E_e may also vary in a different experiment.

For example, the maximum at 6.3 eV in the energy difference curves of the secondary electrons excited from LiF film by soft x-ray has been identified (Henke et al. 1979) as the result of a plasmon decay using (1) with the following values of the parameters: $\hbar\omega_p = 25.3$, $E_g = 13.6$, $\Delta E_v = 3.7$, $\chi = 1$ eV. In addition to the 6.3 eV maximum a weak feature at 11 eV was also observed, which was suggested to be related to the maximum of the unoccupied density of states.

In the experiments performed by Gusarov and Murashov (Gusarov and Murashov 1994) a broad maximum at $E_s = 6$ eV has been observed in the energy difference curves of secondary electrons emitted from a single crystal of LiF by primary electrons with energy from 50 to 210 eV. For primary energies above 60 eV an additional structure near 11 eV emerged.

The energy difference curves of secondary electrons excited by primary electrons with energy 35 eV from a LiF film show similar structure with two maxima at 8.3 and 11.9 eV (Samarin et al. 2004b). In this context for the observation of secondary electron emission from an atomically flat LiF(001) surface irradiated by fast (0.5 meV) grazing incidence protons, Kimura argued (Kimura et al. 2000) that surface plasmons excited by protons are converted into electron-hole pairs with a conversion rate close to 100%.

So, the studies on the secondary emission spectra of LiF can be summarized roughly as follows. Two features are often observed in the energy distribution of

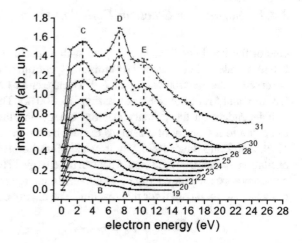

Fig. 2.38 Secondary emission spectra excited from LiF film by primary electrons of various energies. (Reprinted from Samarin et al. (2003) with the permission of AIP Publishing)

secondary electrons: the first—in the energy range between 6 to 8 eV and the second in the energy range between 10 eV to 12 eV. See, for example, the spectra recorded from a LiF film at various primary energies in Fig. 2.38. They do not depend on the type and energy of the primary particles. Possible explanations of these peaks offered in the literature are: (1) manifestation of maxima in the density distribution of unoccupied electronic states; (2) plasmon decay via electron emission from the valence band; (3) autoionization processes. The secondary electron spectra also show the peaks at ~7 and ~11 eV (Samarin et al. 2003). The links between energy loss structures and emission features can be established using two-electron coincidence spectroscopy. For interpretation of these emission maxima one can consider the plasmon decay model. Alternatively the appearance of these peaks might be due to the decay of the excitonic states (Rohlfing and Louie 1998; Caliebe et al. 2000).

2.4.2 Plasmons in Dielectrics

Plasmons are longitudinal collective excitations of electron-hole pairs. They can be excited by incident electrons and show up as a maximum in the energy loss function $Im\left(-\frac{1}{\varepsilon}\right)$. When damping is small the position of this peak coincides with the zero of the dielectric function $\varepsilon\left(\omega_p\right) = 0$.

In metals, electron and hole belong both to the conduction band, and the plasmon consists of intraband transitions. For simple metals, the plasma frequency is given by the free-electron formula:

$$\omega_f^2 = \frac{ne^2}{m\varepsilon_0},$$

(2.25)

where n is the density of electrons and m is the bare electron mass. In undoped semiconductors and insulators, the valence electrons have to be excited across the band gap E_g, and therefore interband transitions constitute the plasmon, which are called sometimes "valence plasmon".

The situation is different in insulators where the exchange interaction is dominant and leads to the formation of excitons. Also, the band structure can no longer be approximated by a nearly free electron model. The first correction to (2) is due to Horie (1959) who derived the following formula:

$$\left(\hbar\omega_g\right)^2 = \left(\hbar\omega_f\right)^2 + E_g^2, \tag{2.26}$$

where E_g is the band gap and ω_f is the free electron plasma frequency of the valence electrons.

In (Egri 1982) the following expression for the plasma frequency of non-metals was derived:

$$(\hbar\omega_x)^2 = \left(\hbar\omega_f\right)^2 + E_x^2, \tag{2.27}$$

where E_x is the energy of the lowest transverse exciton. The derivation is based on the microscopic dielectric function of an excitonic system (Ewing and Seitz 1936; Grimley 1958). It is modelled by two discrete lines situated at E_x and E_g. The total oscillator strength is fixed by the f-sum rule, and the relative strengths are determined according to the system under consideration. In insulators, where the Frenkel exciton is the appropriate model, the oscillator strength is concentrated on E_x, whereas in the Wannier model, applicable for semiconductors, the dominant transition is at E_g. Assuming that $(E_g - E_x)\langle\langle\hbar\omega_f$ in the case of Wannier model, one can replace E_g by E_x in determining the plasmon frequency with confidence.

However, in general, one can expect two, close to each other, plasmon frequencies ω_x and ω_g given by (2.25) and (2.27). For LiF the energies of $\hbar\omega_x$ and $\hbar\omega_g$ can be calculated. First of all, using 6 electrons per LiF molecule one can calculate $\hbar\omega_f \approx 22.5\,\text{eV}$. Then assuming excitonic level $E_x = 10\,\text{eV}$ and band gap $E_g = 13\,\text{eV}$ the plasmon energies are:

$$\hbar\omega_x \approx \sqrt{(22.5)^2 + (10)^2} \approx 24.6\,\text{eV} \tag{2.28}$$

$$\hbar\omega_g \approx \sqrt{(22.5)^2 + (13)^2} \approx 26\,\text{eV} \tag{2.29}$$

2.4.3 Plasmon-Assisted (e,2e) Scattering on LiF Films

To reveal the origin of emission features in the energy difference curves of secondary electrons from LiF films, the (e,2e) spectroscopy with low-energy primary electrons was applied to the LiF film deposited on Si(001) surface (Samarin et al. 2004a Surf

Sci). A set of two-dimensional energy distributions of correlated electron pairs are shown in Fig. 2.39. The numbers on the spectra denote the primary electron energy. Each spectrum is symmetric with respect to the diagonal between the E_1 and E_2 axes because of the symmetry of the experiment with respect to the diagonal of the two electron detectors. The white dashed lines indicate maxima that have the same projection on the E_1 axes in all presented spectra. It means that one electron of the pair in the maximum is preferentially emitted with the same energy of about 7.5 eV independently of the incident electron energy. To underline this finding Fig. 2.40 shows the projections onto the E_1 axis of the two-dimensional energy distributions, which are taken at various primary energies in the range from 22.3 to 42.3 eV. It is seen that at primary energy of about 25 eV the maximum at 7.3 ± 0.3 eV starts to appear and it becomes more prominent at higher primary energies and moves toward lower energy for $E_p > 38$ eV. The second maximum at 10.9 ± 0.3 eV appears at about 30 eV of primary energy. Now consider the maximum in the 2D spectrum, where one electron has energy 7.3 ± 0.3 eV. The energies of both correlated electrons from this maximum are presented in Fig. 2.40 as a function of the incident electron energy in curves B and C. The binding energy of the excited valence electron (A), as well as the energy lost by the incident electron (D), is presented also. These results show that the energy loss of 23 ± 0.3 eV causes the ejection of the electron with energy (7.3 ± 0.3) eV from the valence band at binding energy of (16 ± 0.3) eV.

Now consider the maximum in the (e,2e) spectra where one of the electrons has an energy of 10.9 eV. The analysis of the (e,2e) spectra taken for various primary energies (Fig. 2.41) is done in the same way as for the 7.3 eV maximum. Figure 2.42 shows the result of such analysis.

The ejected electron (B) always has energy of (10.9 ± 0.3) eV, the energy lost by the primary electron (C) is (26.8 ± 0.4) eV and the binding energy (D) of the valence electron is (15.8 ± 0.3) eV.

The above (e,2e) experimental results can be summarized as follows:

(1) The electron energy loss of (23 ± 0.3) eV is "responsible" for the electron ejection with the energy of (7.3 ± 0.3) eV. In other words, the ejection of the electron with energy of (7.3 ± 0.3) eV occurs due to the decay of the collective excitation with energy of (23.3 ± 0.3) eV. See the energy diagram on Fig. 2.43. The nature of energy losses at 23 eV and 26.8 eV is questionable. Fields et al. measured electron energy loss spectra on LiF(100) and LiF(110) as a function of momentum transfer (Fields et al. 1977). On both surfaces they observed energy losses at 23 and 26 eV for large momentum transfer (>1 Å^{-1}). When momentum transfer decreases those two energy losses merge to 25 eV maximum, which corresponds to plasmon excitation in LiF. In our experiment with low energy primary electrons a large momentum transfer for energy losses of 23 and 26.8 eV is expected. Indeed, assuming for the initial and final states of the incident electron the free electron-like parabolic dispersion, the energy loss of 24 eV by 30 eV primary electrons would correspond to the momentum transfer of 1.5Å^{-1}, e.g. in the vicinity of the LiF Brillouin zone boundary.

LiF film on Si(100) crystal

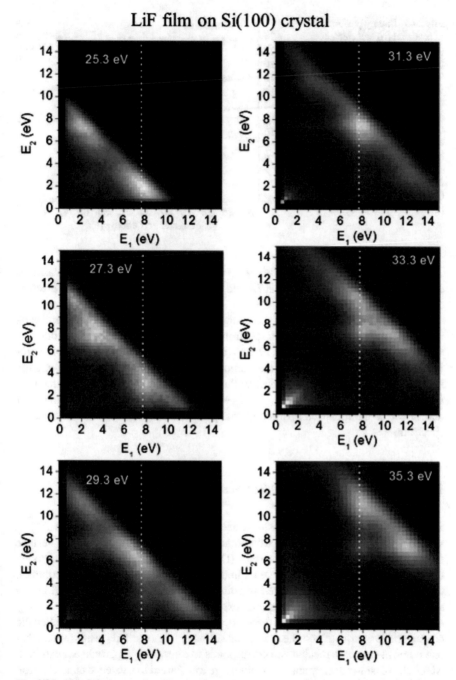

Fig. 2.39 Set of 2D energy distributions of correlated electron pairs excited by incident electrons from a LiF film. Numbers indicate the incident electron energy. Dashed white lines mark maxima in the distributions with the same projections on the E_1 axis. (Reprinted from Samarin et al. (2004a), Copyright (2004), with permission from Elsevier)

Fig. 2.40 Energy positions of scattered (**b**) and ejected (**c**) electrons, as well as incident electron energy loss (**d**) and binding energy of valence electron (**a**) for the scattering reaction in the maxima of 2D distributions (marked in Fig. 2.39 by white lines) presented as a function of the incident electron energy. (Reprinted from Samarin et al. (2004a), Copyright (2004), with permission from Elsevier)

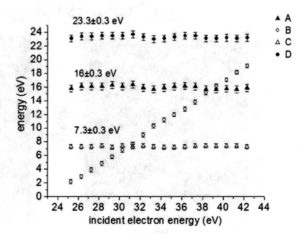

So, if the observed energy losses at 23 and 26.8 eV correspond to the plasmons excitations, the emission features at 7.3 and 10.9 eV are due to their decay via electron ejection (Fig. 2.43).

An alternative explanation of the energy losses and, consequently, of emission maxima in the secondary emission spectrum of LiF film, could be the following. The excitonic states at about 10.6 and 12.5 eV are populated upon the impact of the projectile electron and subsequently the excitonic states auto-ionize via exciton-exciton interaction. The energy with respect to the vacuum level of the resulting electrons will then be localized around $(E_{ex} - (E_g - E_{ex}) - \chi)$ which gives approximately 7 and 11 eV as the localized position of this peaks in the secondary electron energy spectrum.

2.4.4 Plasmon-Assisted (e,2e) Scattering on Metals

Similar effects of a plasmon-assisted secondary electron emission from metals were observed using the two-electron coincidence technique on an Al(100) surface (Werner et al. 2008), a Be(0001) surface (Di Filippo et al. 2016), and on a polycrystalline Al surface (Werner et al. 2011). In these experiments the energy distribution of secondary electrons is measured in coincidence with the energy loss ΔE corresponding to the surface or bulk plasmon excitation. A large increase in the secondary electron yield was observed when the energy loss of the primary electron equalled the characteristic energy of volume (bulk) and surface plasmons. This result reveals that one relevant emission mechanism corresponds to direct single-particle excitation in which the plasmon energy and momentum are transferred to a valence band electron of the solid. The coupling between the direct ionization channel and the plasmon-assisted channel results in a resonant increase of the secondary emission (Di Filippo et al. 2016). Kouzakov and Berakdar (2012) proposed the ejection of a secondary

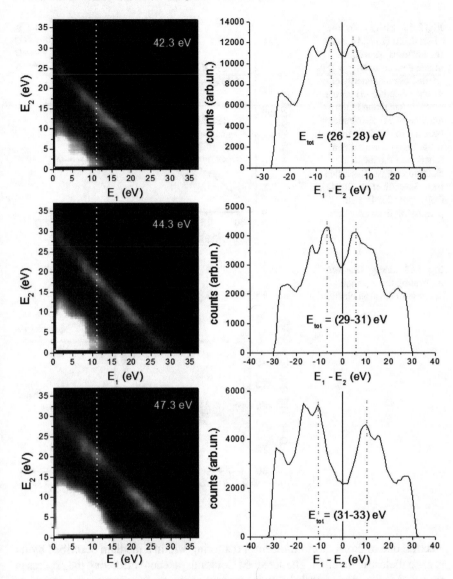

Fig. 2.41 2D energy distributions of correlated electron pairs from LiF film measured at various primary energies (left column) and energy sharing distributions along the ridges of the corresponding 2D spectra (right column). The white dashed line in the left column marks the maxima with one electron of the pair having energy of 10.9 eV. Dashed lines in the right column show corresponding maxima in the sharing distributions. (Reprinted from Samarin et al. (2004a), Copyright (2004), with permission from Elsevier)

Fig. 2.42 Energy positions of scattered (**b**) and ejected (**c**) electrons, as well as incident electron energy loss (**d**) and binding energy of valence electron (**a**) for the scattering reaction in the maxima of 2D distributions (marked in Fig. 2.41 by white line) presented as a function of the incident electron energy. (Reprinted from Samarin et al. (2004a), Copyright (2004), with permission from Elsevier)

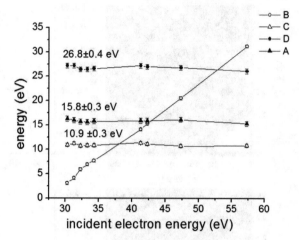

Fig. 2.43 Energy diagram illustrating the plasmon decay with electron ejection

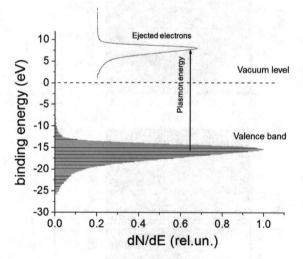

electron due to direct electron-electron scattering within a medium described by its inverse dielectric function. The screened Coulomb potential mediating the process is such that the scattering probability is resonant at energy transfers corresponding to the excitation of collective modes, such as volume or surface plasmons. The model predicts the general features (Kouzakov and Berakdar 2012) of the cross section measured on Al (Werner et al. 2008) but, lacking the solid band structure, it fails to describe fine details.

2.5 Application of (e,2e) Spectroscopy for Studying Surfaces and Surface Phenomena

2.5.1 Measurements of Energy Band Parameters of a Dielectric Surface

In a simplified but very practical manner dielectrics are characterized by an energy diagram that includes few parameters such as: electron affinity χ, bandgap E_g, width of the valence band Δ, energy position of an exciton level (band) E_{ex} (see Fig. 2.44). These parameters are very important for understanding of electrical and optical properties of dielectrics and building electronic and electro-optical devices based on these materials.

Experimentally determined band parameters are used to verify ab initio energy band structure calculations, self-consistent approaches and empirical calculations for these materials. On the other hand there is no experimental technique at present, which provides all mentioned above parameters. Usually it is necessary to combine the data of various techniques such as ultraviolet photoelectron spectroscopy (UPS), electron energy loss spectroscopy (EELS), x-ray photoelectron spectroscopy (XPS), photoabsorption and reflection measurements, two-photon absorption, recombination luminescence, photoconductivity and others in order to arrive at a complete picture (Poole et al. 1975). Results of the measurements depend very much on experimental technique applied and on the sample conditions, such as surface charging (Schlaf et al. 1998), sample preparation (Lapiano-Smith et al. 1991) and its temperature (Roy et al. 1985), surface contaminations (Singh and Gallon 1984) and surface degradation during measurements (Golek and Sobolewski 1999). For example, the Ultraviolet Photoelectron Spectroscopy (UPS) measurements shows negative electron affinity for the epitaxial layer of LiF on Ge(100), whereas the LiF film on Si(100) surface with a substantial mismatch does not exhibit a negative affinity (Lapiano-Smith et al. 1991). For a LiF layer (about 30 nm thickness), studied with XPS, an affinity of 1.8 eV was determined using experimental data and a "best fit" to the model calculations (Henke et al. 1979). Measured band gap of LiF is ranging from

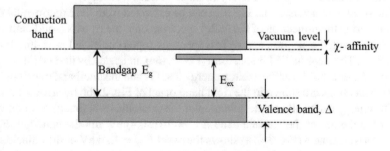

Fig. 2.44 Basic energy band parameters of a dielectric

12 eV (Guo et al. 2000) to 14.4 eV (Golek 1993). The measured valence bandwidth of LiF varies from 4 eV (Morgner 1999) to 6.1 eV (Lapiano-Smith et al. 1991). Therefore it is desirable to have a technique or combination of techniques in situ for the consistent measurements of the energy band parameters of a dielectric. These technique must be nondestructive and with a little charging effect. It was demonstrated (Samarin et al. 2004b Solid State Comm) that a combination of the conventional one-electron spectroscopy (EELS) and the two-electron spectroscopy ((e,2e)) can be used for measuring basic parameters of an insulator. As an example the results for a LiF film deposited on Si(001) surface are presented here. In this example the LiF film was deposited on clean Si(001) surface by thermal evaporation from a molybdenum crucible heated by electron bombardment. Before placing into the vacuum the silicon wafer was chemically cleaned and in vacuum it was outgassed and heated up to 1250 °C until low energy electron diffraction patterns from reconstructed Si(100) (2 × 1) surface appeared. A combination of SEM and TEM has been used to determine the structural and compositional properties of the LiF films grown on Si. SEM imaging has been applied to study the morphology of the LiF films. Structural and compositional properties of the films have been studied by high-resolution imaging, selected area diffraction and energy-filtered imaging using a JEOL 3000F FEGTEM. The diffraction patterns obtained from a plan view sample at the Si(001) zone axis show extra set of diffraction spots identified as 220 reflections from the LiF crystals in addition to those from the Si substrate. In addition, {111}-lattice spacing of LiF has been measured from cross-sectional sample by high resolution TEM.

Energy-filtered images show contrast associated with the LiF crystallites. This contrast is maximized at an energy loss of ~ 26 eV that is close to the characteristic energy loss of the LiF plasmon (25.3 eV Henke et al. 1979). TEM studies have shown that the LiF films are crystalline with crystallite sizes in the range 3–10 nm. The film thickness was estimated using transmission electron microscopy and ellipsometry to be about 100–150 Å.

The electron energy loss spectra were measured on LiF film excited by 25–50 eV primary electrons. The (e,2e) spectra were measured on the same LiF film using the same primary electron energy. Using Total Current Spectroscopy (TCS) (Komolov 1992) it was found that there is a contact potential difference of 2.3 eV between the deposited film and the filament of the electron source (see Fig. 2.45). This contact potential difference allows to correct an impact electron energy.

The band parameters of the LiF film can be extracted from the experimental data by comparing the projection of the (e,2e) spectrum onto the E_1 axis (solid circles on Fig. 2.46a) with the electron energy loss spectrum (open circles in Fig. 2.46a) for E_p = 28.3 eV. The steps on EELS spectrum (Fig. 2.46a), indicated by dashed lines 1 and 2, show the thresholds for the exciton energy loss (1) and the interband transition (2). These excitations are shown in the right hand panel of Fig. 2.46b by arrows 1 and 2.

The energy losses allow determination of the excitonic level (band) position with respect to the top of the valence band $E_{ex} = (10 \pm 0.4)$ eV and the band gap $E_g = (13 \pm 0.4)$ eV. Line 3 (Fig. 2.46a) shows the onset $E_{coi} = 13.6$ eV of the coincidence spectrum projection. It is clear that the onset of the coincidence spectrum projection is shifted towards lower energies with respect to the onset of the interband transition

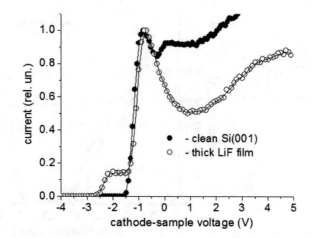

Fig. 2.45 Total current spectra of clean Si(001) surface (bold circles) and thick LiF film on Si(001) (open circles)

in the EELS because in the coincidence spectrum the electron is detected only if the second electron of the pair is detected with the energy E_2 (Fig. 2.46b, left panel). The onset (maximum energy of the first electron) corresponds to the minimum energy of the second electron due to the energy conservation law. Taking into account that the minimum electron energy detected in the coincidence experiment is 0.7 eV (determined from the two-dimensional (e,2e) spectrum) one can calculate the electron affinity χ of the actual LiF film: $\chi = E_p - E_g - E_{coi} - E_2 = (1 \pm 0.4)$ eV. The valence band width of the LiF film is estimated using the binding energy spectrum measured with 22.3 eV primary electron energy (Fig. 2.47).

The band width at half maximum is (3.1 ± 0.3) eV. Using linear interpolation on both sides of the curve (Fig. 2.47) the full width of the valence band is estimated to be equal to $\Delta = (6 \pm 0.5)$ eV. We would like to point out that the measured energy band parameters of the LiF film characterize the particular film of 15 nm thickness deposited on the Si(001) substrate.

The film morphology is shown in the secondary electron image of Fig. 2.48. The image is taken with Field Emission Scanning Electron Microscope (FESEM).

The results of the measurements are consistent with published data. The measured band gap energy of $E_g = (13.0 \pm 0.4)$ eV and the valence bandwidth of $\Delta = (6.0 \pm 0.5)$ eV are close to the values presented in (Poole et al. 1975): 13.6 and 6.1 eV, respectively. On the other hand the affinity measured in our experiment $\chi = (1.0 \pm 0.4)$ eV is quite different from the negative value reported in (Lapiano-Smith et al. 1991) $\chi = -2.7$ eV. This difference most likely is due to the fact that in (Lapiano-Smith et al. 1991) the measurements were performed on epitaxial layer of LiF whereas in our case the film was far from epitaxial as shown in Fig. 2.48. The advantages of the described technique are: (i) very low incident electron current and (ii) low electron energy that minimizes the sample surface charging and its destruction under the measurements. All parameters are measured using the same experimental setup providing self-consistence of the measurements.

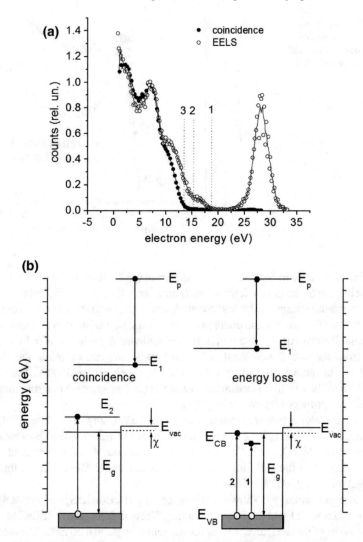

Fig. 2.46 a Electron energy loss spectrum (EELS) for a LiF film measured at $E_p = 28.3$ eV (open circles) and the projection of the 2D (e,2e) energy distribution (solid circles) on the E_1 axis measured for the same film at the same primary energy; **b** energy diagrams of the (e,2e) (coincidence) reaction and energy loss reaction; in the first case two electrons with energies E_1 and E_2 are detected, whereas in the second case only one electron with energy E_1 is detected. (Reprinted from Samarin et al. (2004b), Copyright (2004), with permission from Elsevier)

Fig. 2.47 Total energy distribution (binding energy spectrum) of correlated electron pairs excited from the LiF film by electrons with $E_\mathrm{p} = 20.3$ eV. (Reprinted from Samarin et al. (2004b), Copyright (2004), with permission from Elsevier)

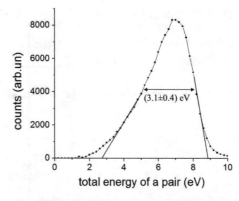

Fig. 2.48 Morphology of the LiF film. (Reprinted from Samarin et al. (2004b), Copyright (2004), with permission from Elsevier)

100nm

2.5.2 Two-Electron Distributions from Various Materials Excited by Electron Impact

To demonstrate how electronic and geometric structure influence the (e,2e) distributions excited by incident electrons we present a brief overview of the (e,2e) spectra recorded on different samples.

2.5.2.1 GaAs (001) Crystal

Using (e,2e) spectroscopy a GaAs single crystal was studied as an example of a semiconductor. One of the interesting features of the GaAs band structure is the presence of the "Light holes" and "Heavy holes" energy bands at the top of the valence

Fig. 2.49 Total energy
distribution and LEED
patterns of GaAs(001)
surface

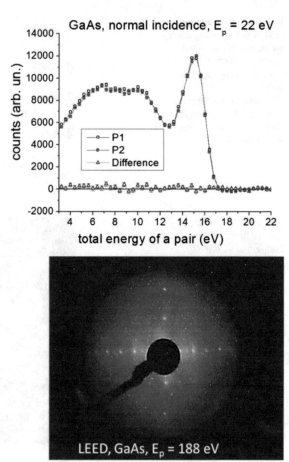

band. In addition there is a split-off energy band due to the spin-orbit interaction. The question is how these features show up in the (e,2e) spectra, and whether there is a spin-dependence in the (e,2e) scattering. The surface was cleaned in the vacuum by ion spattering and annealing prior to the measurements. The bottom panel of Fig. 2.49 shows image of the LEED patterns taken at 188 eV energy of electron. Figure 2.49. (top panel) represents the total energy distribution of correlated electron pairs excited from GaAs (001) surface by spin-polarized electron impact with electron energy of 22 eV. Well defined maximum at about 15 eV of total energy indicates a substantial density of states at the top of the valence band. No spin dependence was found as evidenced from total energy distributions.

Fig. 2.50 Total energy distributions of correlated electron pairs from clean Gd layer and Gd layer with adsorbed oxygen in the range from 0 to 1 monolayer (ML)

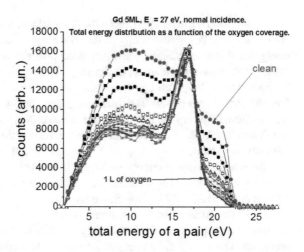

2.5.2.2   Gd Layer on W(110)

Gadolinium layer on W(110) was studied using spin-polarized (e,2e) spectroscopy in view to observe a possible surface ferromagnetic layer of Gd at room temperature. In addition, a surface electronic structure transformation upon oxygen adsorption was also in the focus of the study since Gd is a very active metal. The Gd layer was deposited on W(110) using thermal evaporation from Mo crucible (EFM-3, OMICRON). The 5 monolayer Gd film on W(001) surface was excited by 27 eV spin-polarized incident electron beam. The (e,2e) spectra were measured at room temperature. No spin-effect was observed in the (e,2e) spectra indicating that the top most layer of Gd is not ferromagnetic at room temperature. Figure 2.50 shows total energy distributions of the correlated electron pairs excited from clean 5 atomic layers Gd film on W(001) and the same film with progressively increased adsorbed layer of the oxygen, from 0 to 1 monolayer (ML). The amount of oxygen on the Gd surface was increased by steps of 0.1 ML. It is seen in Fig. 2.50 that the adsorption of 0.1 ML of oxygen changes dramatically the total energy distribution: a maximum at about 16.5 eV of total energy (6 eV below the Fermi level) appeared. This maximum is associated with the $2p$ orbital of oxygen that participates in the oxygen-metal bonding. The maximum increases with increase of the oxygen coverage and becomes a dominant feature of the spectrum at about 0.7 ML. At the coverage of 0.8 ML an additional maximum at total energy of about 11.5 eV shows up indicating further transformation of the electronic structure of the gadolinium oxide.

2.5.2.3   MgO Crystal

Crystal of MgO has found numerous applications in solid state electronics, optical devices, and as a substrate for deposition of various metal films.

Any additional information about its electronic structure is of great value. Application of the (e,2e) spectroscopy to the MgO crystal can provide parameters of its energy band structure. Since it is a very good insulator it was heated to about 100 °C during the measurements to avoid charging. The crystal was cleaned up by ion spattering and annealing until the sharp diffraction patterns appear on the LEED screen (see Fig. 2.51, lower panel). The total energy distributions of correlated electron pairs excited from MgO crystal by spin-polarized electrons with energy from 23 to 30 eV are shown in Fig. 2.51 (upper panel). Spectra recorded with spin-up and spin-down electrons are identical, i.e. no spin effect was observed. The onset of the total energy distribution for primary energy of 27 eV is located at about 18 eV. It means that the energy difference between the top of the MgO valence band and the vacuum level is equal to $27 - 18 = 9$ eV. Since the energy gap of MgO is $E_g = 7.83$ eV the affinity of the surface can be calculated: $\chi = 9 - 7.83 = 1.17$ eV. When the incident electron energy increases the low-energy maximum of the total energy distributions increases. This maximum corresponds to the multi-step (at least two-step) electrons excitations.

2.5.2.4   Thin Layers of LiF and Ag on Si(001) Crystal

Silicon is the major semiconductor used in the solid state electronics today. Silicon wafers are used as substrates for multilayered structures, which forms electronic devices. Thin layers of dielectrics, metals, and semiconductors are usually grown on a wafer by a variety of techniques. Electronic properties of such systems are in the focus of applied research.

The Si(001) crystal and thin layers of LiF and Ag on the surface of this crystal are chosen as samples for testing potential application of the (e,2e) spectroscopy for studying various solid state structures. The Si(001) was cleaned inside the vacuum chamber by resistive heating up to 1600 K. Appearance of the sharp diffraction patterns indicated a removal of the oxide layer from the surface of the Si(001) sample. For example, at 25 eV primary energy the ratio of the intensity of the (00) diffraction pattern to the surrounding background was about 25. Figure 2.52 shows the total energy distribution of correlated electron pairs excited from Si(001) by 27 eV incident electrons (bold squares). The bump at 14.5 eV (shown by arrow) probably indicates a presence of the remaining oxygen. A deposition of the (15 ± 5) nm LiF film on the Si (001) surface changes dramatically the total energy distribution (or binding energy spectrum). Indeed the onset of the spectrum corresponding to the pairs excited from the top most occupied levels (top of the valence band) is shifted from 20.5 to 13 eV. This indicates the relative position of the top edges of the valence bands of the Si and the LiF film. The substantial drop of the number of pairs with total energy below 8 eV in the case of LiF film is explained by the presence of the large (13 eV) band gap in LiF film. Indeed this band gap does not allow for secondary electrons to undergo another inelastic scattering with excitation of a valence electron if the energy of the secondary electron is below 13 eV relative to the bottom of the conduction band (which almost coincides with the vacuum level). Therefore, at relatively low primary energy the multi-step scattering events in the LiF film are suppressed that leads to the deficit of low-energy pairs. An energy diagram illustrating difference between a dielectric and a metal in relation to the multi-step contributions to the (e,2e) spectrum is shown in Fig. 2.53.

A metal layer on a Si surface usually serves as an electric contact to the structure. The quality of the layer and interface between metal and the semiconductor in many cases determines the functionality of the device. Silver or gold layers are often used as such contacts. Besides, an array of Ag of Au nanoparticles on a silicon surface may be used as a substrate for very sensitive molecular analysis using Plasmon-Enhanced Raman Spectroscopy (Michaels et al. 1999). Presence of thin metal layers on a semiconductor surface changes optical properties of the surface. So, the application of the (e,2e) spectroscopy to a metal-semiconductor structure is an obvious choice to test analytical power of this technique. Figure 2.54 represents total energy distributions (upper panel) and energy sharing distributions (lower panel) of correlated electron pairs excited by 25 eV primary electrons from Si(001) surface and thin layers of Ag on Si(001). One can see that the deposition of a silver layer changes both total energy distribution and the energy sharing distribution. In the total energy distribution large maximum at 8.5 eV is moved to the position at 14.5 eV. At a certain thickness of the Ag film the maximum at 18.5 eV corresponding to the excitation of electrons from the Si valence band is moved to the position at 20 eV corresponding to the electrons excitation from the vicinity of the Ag Fermi level (show by arrow A in the Fig. 2.54).

In the energy sharing distribution the deposition of Ag film leads to the relative decrease of the probability of excitation of pairs with equal energies and increase of the probability of the pairs excitation with very unequal energies (see bottom panel of Fig. 2.54).

2.5.2.5 Comparison of Binding Energy Spectra with Densities of States

Au(111) and Cu (111) Comparison

Number of correlated electron pairs as a function of the total energy of a pair is determined by the number of appropriate initial (occupied) and final (unoccupied) states, matrix element of corresponding transitions, and available phase space corresponding to the detectors' acceptance angles and energy resolution. However, one can expect that the binding energy distributions will reflect qualitatively the distribution of density of states (electrons can be excited only from states with non-zero density of states). If the pairs are selected with almost equal energies (or equal momenta) then the distribution would characterize the "density of states" in the centre of the Brillouin zone. Figure 2.55 shows such distributions for Au(111) and Cu(111). The pairs are selected under condition that $(E_1 - E_2) < 0.05 \ (E_1 + E_2)$. It is seen from Fig. 2.55 that the distributions are different. Both distributions have a maximum just below zero binding energy (position of the Fermi level). This maximum corresponds to the excitation of pairs from Shockley-type surface states (Samarin et al. 2015

Fig. 2.51 Total energy distribution of correlated electron pairs excited from MgO by 27 eV electron impact (top panel); LEED patterns (bottom panel)

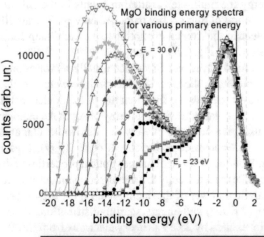

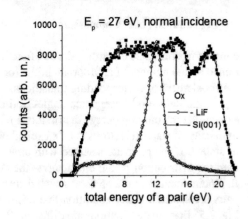

Fig. 2.52 Total energy distributions of correlated electron pairs excited by 27 eV electrons from Si(001) surface and thin LiF film on Si(001)

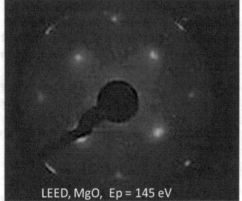

Fig. 2.53 Energy diagram
illustrating difference of the
(e,2e) scattering in a
dielectric (right) and a metal
(left)

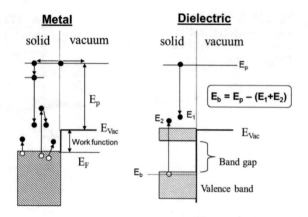

Fig. 2.54 Total energy
distributions (top panel) and
energy sharing distributions
of correlated electron pairs
excited by 25 eV electrons
from Si(001) and Si (001)
covered by Ag layers

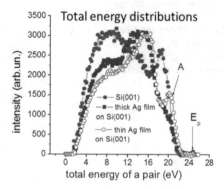

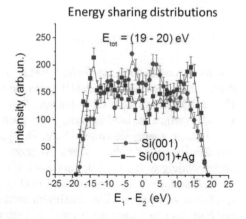

JESRP). Maxima related to the pair excitation from *d*-derived states are located at
different energies in these two materials.

Fig. 2.55 Binding energy spectra of Au(111) and Cu(111) in the centre of the Brillouin zone

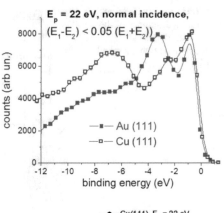

Fig. 2.56 Binding energy spectra of Cu(111) and Cu(001). (Reprinted from Samarin et al. (2015), Copyright (2015), with permission from Elsevier)

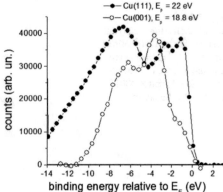

Cu(111) and Cu(001) Comparison

It is interesting to compare binding energy spectra of Cu(111) and Cu(001) because the first one possesses surface states, but the second one does not have surface state. Figure 2.56 shows such a comparison. Indeed, Fig. 2.56 clearly indicates that for Cu(001) surface the binding energy spectrum shows low intensity in the range from zero binding energy down to −2 eV. This part of the spectrum corresponds to the low density *s-p* hybridized states. At about −4 eV binding energy the maximum corresponding to the high density *d*—derived states shows up. In contrast to the Cu(001) surface the spectrum of the Cu(111) surface shows high intensity maximum just below zero energy. This maximum corresponds to the Shockley surface states. So, both spectra shows three maxima with different relative intensity that reflex different electronic structure of these surfaces.

Cu(001) and Ni (001) Comparison

A comparison of binding energy spectra of Cu(001) and Ni(001) can demonstrate to what extent the measured spectra reflect densities of states. Indeed, the Cu(001) surface has low density of s-p hybridized states just below the Fermi level (in the 2 eV energy band). In contrast, the Ni(001) surface has high density d-states crossing the Fermi level. This difference in the density of states distributions should show up in the measured binding energy spectra. Figure 2.57 represents the comparison of the measured binding energy spectra of Cu(001) and Ni(001) (background is not subtracted) and corresponding densities of states. One can see that for Cu(001) a low intensity shoulder extends from 0 down to -2 eV binding energy and then a pronounced maximum appears at about -3.5 eV binding energy. The calculated DOS well reproduces the measured spectrum in the range $(0 \div -6)$ eV, except the triple peak of d-states, this is not resolved in the measured spectrum. In the case of Ni(001) sample the onset of the measured binding energy spectrum (zero binding energy or the Fermi level position) crosses high density d-states. Again the triple maximum of the calculated DOS is not resolve in the measurements, but the position of the averaged maximum is well reproduced by the measurements.

2.6 Mechanism of (e,2e) Reaction in Metals, Semiconductors and Dielectrics

As pointed out earlier for a crystal sample the parallel to the surface component of the electron momentum is conserved. Therefore the off-normal incidence in the (e,2e) scattering on a crystal surface leads to an asymmetric energy sharing distribution with respect to the point where $E_1 = E_2$. However, it seems that it is perfectly true only in the case of a metal crystal samples. Figure 2.58 shows an example of energy sharing distributions measured at the same geometrical arrangement on different crystal samples: Si(001), MgO(001), W(110).

One can see that for Si(001) and MgO(001) crystals the energy sharing distributions are symmetric with respect to the point where $E_1 = E_2$ for normal incidence as well as for off-normal incidence. On the other hand the energy sharing distribution of correlated electron pairs excited from W(110) with off-normal incidence of electrons is quite asymmetric with respect to the point where energies of both electrons are equal.

A very interesting transformation of the energy sharing distribution of electron pairs from W(110) excited by electron impact is observed when oxygen atoms are adsorbed on the surface. Figure 2.59 represents binding energy spectra and energy sharing distributions of correlated electron pairs excited from a clean W(110) surface and from a W(110) surface covered by ½ of monolayer and 1 monolayer of oxygen. The oxygen coverage is confirmed by LEED patterns shown in the right column of Fig. 2.59. The binding energy spectrum of W(110) with half of a monolayer of

Fig. 2.57 Binding energy
spectra of Cu(001) and
Ni(001) excited by 18.8 eV
primary electrons at normal
incidence. Background is not
subtracted. Measured spectra
compared with densities of
states (DOS)

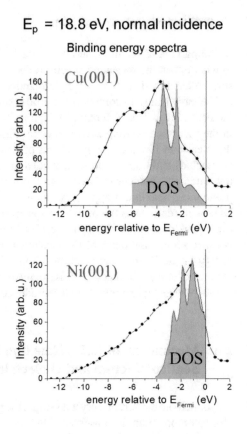

oxygen (Fig. 2.59c) contains a characteristic peak at a binding energy of about 6 eV
relative to the Fermi level. This peak corresponds to the electron pair excitation from
the *O2p*-derived states resulting from a hybridization of *2p* states of an oxygen atom
and valence states of the tungsten crystal. At this coverage only half of the W(110)
surface is covered by oxygen atoms, and this is enough to change substantially the
energy sharing distribution of electron pairs excited from the vicinity of the Fermi
level (Fig. 2.59d). It is still asymmetric with respect to the $E_1 = E_2$ point of the energy
sharing distribution, but the asymmetry is much less pronounced than in the case of
a clean W(110) surface. When oxygen atoms form a monolayer on the surface of the
tungsten crystal with the structure O(1 × 1), as evidenced from the corresponding
LEED picture, the oxygen maximum on the binding energy spectrum becomes more
pronounced (Fig. 2.59c) and the asymmetry in the energy sharing distribution of
electron pairs disappears (Fig. 2.59f). Taking into account that the (e,2e) spectroscopy
is very surface sensitive one can assume that in the case corresponding to Fig. 2.59e,
f the correlated electron pairs are generated in the surface region of the sample
consisting of a layer of tungsten atoms and a layer of oxygen atoms.

   Thus, the presented above results indicate that the mechanism of the correlated
electron pairs generation by an electron impact for the case of a metal crystal is

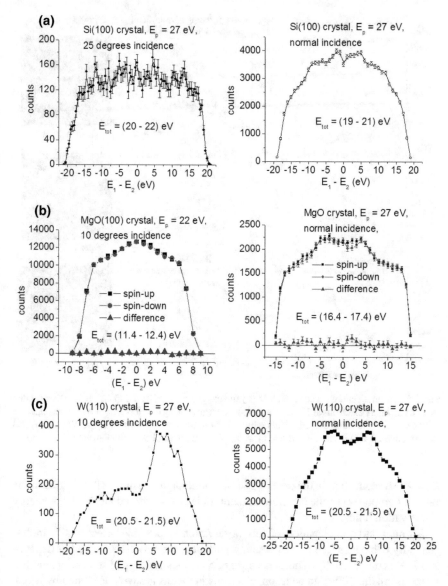

Fig. 2.58 A set of energy sharing curves of various single crystals for a normal and off-normal incidence: **a** Si(100), **b** MgO(100), **c** W(110)

different from that of a semiconductor and insulator crystals. This difference shows up in the different symmetry of the energy sharing distributions of electron pairs excited from the top of the occupied electronic states of the crystals at off-normal incidence. In the case of a metal crystal this distribution is asymmetric, but in the case of a semiconductor and insulator it is symmetric with respect to the point where

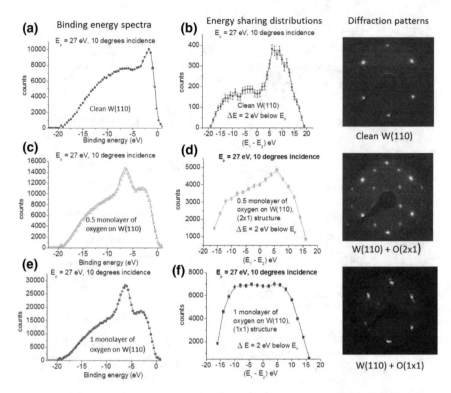

Fig. 2.59 Binding energy spectra of clean (**a**) and oxygen covered (**c** and **e**) W(110) surface excited by 27 eV primary electrons at 10° incident angle; energy sharing distributions of correlated electron pairs excited at the same experimental conditions from the valence band of clean W(110) (**b**) and oxygen covered W(110) surface (**d** and **f**) within $\Delta E = 2$ eV just below the Fermi level

$E_1 = E_2$. It seems that in the case of a semiconductor and insulator the parallel to the surface component of the electron momentum is not conserved although the energy conservation holds.

An alternative explanation of this observation is possible. Electrons in metals generally behave as free non-interacting particles and may be modelled as a gas of non-interacting particles. This is because of a very short range of screening length λ ~ 0.5 (Berakdar 1999). In addition, electrons in metals are well described by Bloch wave functions (extended states) where the parallel-to-the-surface component of the quasi-wave vector is a "good" quantum number (conserved). Therefore the scattering of an incident electron by a valence electron of a metal crystal can be viewed as an interaction of two individual electrons with conservation of their energy and parallel to the surface momentum. At off-normal incidence correlated electron pairs excited by an electron impact carry a parallel to the surface component of the incident electron. Therefore the energy sharing distribution of electron pairs excited from the valence states just below the Fermi level is asymmetric with respect the point where $E_1 = E_2$. In the case of a semiconductor or an insulator the electronic states are more adequately described by Wannier functions (localized states). Therefore scattering of an incident electron by a valence electron of a target in this case resembles more the case of an atomic target. Therefore the momentum of the incident electron in the collision is also transferred to the atomic ion core or to the crystal as a whole, whereas the energy is shared between scattered electron and ejected electron of the solid. This is because the mass of the atomic core (or the crystal) is much larger than the electron mass, therefore the incident electron energy is divided between electrons, but the momentum of the incident electron is absorbed by the ion core. In addition, an obvious difference between metals, on one side, and semiconductors and insulators, on the other side, is the different screening length. Indeed, in tungsten, for example, the Thomas-Fermi screening parameter $\lambda \sim 0.5$ Å, whereas in semiconductors and insulators it might be of the order of microns. Intuitively, one can expect that because of the short screening parameter in metals, the electron-electron scattering is more of an individual electron-electron encounter than in semiconductors or insulators.

An alternative approach to the comparison of (e,2e) scattering on metals, semiconductors and insulators is the following. The main question is whether the scattering occurs in the field of one scattering center (the so-called binary interaction) or in the total field of several (many) scattering centers (the case of multi-particle interaction). The conservation of total energy and momentum is fulfilled in the binary scattering. In this case, the primary electron interacts predominantly with one valence electron and the total momentum of the pair after scattering is equal to the sum of the electron momenta before scattering. However, the condition for the conservation of the total momentum of the electron pair is violated if the scattering of the primary electron occurs in the total field of several valence electrons. The transition from scattering with conservation of the tangential component of the momentum to scattering without conservation can be associated with a change in the character of the interaction from binary to multi-particle.

In the low energy region the nature of the interaction is determined by the ratio of the mean distance between the valence electrons l_d to the screening parameter (or screening length λ). The first parameter is determined by the density of valence electrons for which the definition is the same for metals, semiconductors and dielectrics while the screening length depends on the predominant screening mechanism. Where the screening length is less or much less than the average distance between the valence electrons (metals), the effective Coulomb interaction region of the primary electron with the valence electron is limited to a spherical region of radius λ and the interaction has a binary nature. The potential between these spherical regions is determined by the sum of the "tails" of the screening potentials of the valence electrons. This total potential has little variation, the scattering cross section is much lower than in the region of the central spherical potential and the scattering in this region has a many-particle nature. In the opposite case, when the screening length is of the order of, or greater than, the average distance between valence electrons (as for non-metals), the potential in the region between valence electrons is the sum of the central potentials of all the nearest valence electrons and the scattering has a multiparticle nature. As the relative energy of the interacting electrons increases, the primary electron penetrates deeper into the central field of the chosen valence electron and binary scattering can become the determining scattering mechanism. For example, at an energy of the order of 50 keV (Bowles et al. 2004), the scattering by valence electrons of silicon looks like scattering by free electrons with parabolic dispersion.

We apply below the model of Thomas-Fermi screening with reference to oxygen adsorption on the tungsten crystal (see Fig. 2.59). The density of electronic states (DOS) in the momentum space near the Fermi level defines the screening length in metals. It can be calculated using density functional method (Blaha et al. 2001) and applying topological analysis to the calculated three-dimensional distribution of the electron density.

The experimental results in Fig. 2.59. illustrate the change in the character from binary to multi-particle scattering of electrons on the surface of tungsten upon adsorption of oxygen. It is seen that the atomic structure of the adsorbed layer of oxygen atoms repeats (with a certain lateral shift) the lateral arrangement of tungsten atoms, which is confirmed by the diffraction pattern of slow electrons. Specific calculations of the electronic structure of the adsorbed oxygen layer by the density functional method show the absence of a forbidden band near the Fermi level and the presence of free valence electrons in this region, just as in the case of a layer of tungsten atoms. Of course, the density of the valence electrons and the density of electronic states near the Fermi level will differ for a layer of tungsten atoms and a layer of oxygen atoms. It can be assumed that the change in the nature of the electron scattering during the adsorption of oxygen reflects the change in the density of the valence electrons and the change in the density of electronic states near the Fermi level upon transition from the surface layer of the single crystal of tungsten to the adsorbed layer of oxygen atoms. Consequently, the screening length changes in turn.

2.7   Selection Rules in (e,2e) Scattering from Surfaces

In order to empower the (e,2e) spectroscopy it is essential to understand in detail the collision dynamics and its spin-dependence (see Morozov et al. 2002).

This problem with an emphasis on the symmetry properties of the spatial part of the valence states, which is involved in the (e, 2e) process was addressed in (Feder and Gollisch 2001; Gollisch and Feder 2004), where the spin- and spatial-symmetry-dependent selection rules were obtained. These rules are deduced by analysing the spin-dependent cross section of the (e,2e) reaction on a crystal surface and are summarized as follows (Feder and Gollisch 2001).

2.7.1   First Selection Rule

If the spatial part of the wave function $\varphi_2(r)$ of the valence electron involved in the (e,2e) collision is **antisymmetric** with respect to the scattering plane $R2$, $\varphi_2(x, y, z) = - \varphi_2(x, -y, z)$ (see Fig. 2.60), then the cross-sections $I_{\downarrow\downarrow} = 0$ and $I_{\downarrow\uparrow} = 0$ and for

Fig. 2.60 The (e,2e) scattering set-up with two mirror symmetry planes *R1* and *R2*

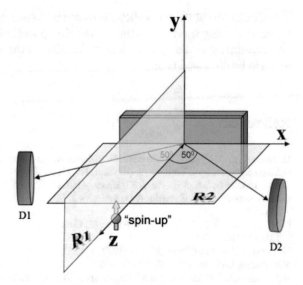

unpolarized incident beam $I_{\downarrow\downarrow} + I_{\downarrow\uparrow} = 0$. Here $I_{\downarrow\downarrow}$ and $I_{\downarrow\uparrow}$ denote the cross-sections of the correlated electron pairs generation with spins of the incident and the valence electrons being parallel and the cross-section of electron pairs generation for anti-parallel spins' orientation, respectively. It means that valence electron states, which are antisymmetric with respect to the reaction plane, do not manifest themselves in (e, 2e) scattering cross sections.

2.7.2 Second Selection Rule

If the valence electron wave function $\varphi_2(\mathbf{r})$ is **symmetric** with respect to the plane *R1* (see Fig. 2.60) $\varphi_2(-x, y, z) = \varphi_2(x, y, z)$, and electrons are detected at equal angles with respect to the surface normal, and $E_1 = E_2$, then the cross-sections $I_{\downarrow\downarrow} = 0$ and $I_{\downarrow\uparrow} \neq 0$. It means that for this experimental arrangement the triplet cross-section vanishes, and only the singlet channel is open. It should be noted that for the same experimental conditions, but for the case when the valence electron wave function is **antisymmetric** with respect to the plane *R1*, i.e. $\varphi_2(-x, y, z) = -\varphi_2(x, y, z)$, both scattering channels are open, i.e. cross-sections $I_{\downarrow\downarrow} \neq 0$ and $I_{\downarrow\uparrow} \neq 0$.

Although these two selection rules are strictly speaking valid only in the absence of spin-orbit coupling and for perfectly well-defined energy, momenta and angles, they will hold in good approximation and be useful over wide energy and momentum ranges, in which the valence electron spinor is dominated by a particular spatial symmetry type (Gollisch and Feder 2004).

The second selection rule has important consequences for the interpretation of equal energy sharing spectra from a ferromagnetic sample in terms of the spin-split valence electron structure. It allows connections between measured spin-asymmetry (calculated using spectra measured with spin-up and spin-down incident electrons) and asymmetry of the spectral density function of the valence electronic structure that will be discussed later.

References

U. Amaldi Jr., A. Egidi, R. Marconero, G. Pizzella, Rev. Sci. Instrum. **40**, 1001 (1969)

O.M. Artamonov, Jour. Tech. Phys. (Russian) **55**, 1190 (1983)

O.M. Artamonov, S.N. Samarin, J. Kirschner, Appl. Phys. A 65 535–42 (1997)

O.M. Artamonov, S.N. Samarin, A.O. Smirnov, J. Electron Spectrosc. Relat. Phenom. **120**(1–3), 11 (2001)

O.M. Artamonov, S.N. Samarin, S. Paolicelli, G. Stefani, J. Electron Spectrosc. Relat. Phenom. **131–132**, 105 (2003)

N.R. Avery, AIP Conf. Proc. **36**, 195 (1977)

N.R. Avery, Surf. Sci. **111**, 358 (1981)

R.Z. Bachrach, F.C. Brown, S.B.M. Hagström, J. Vac. Sci. Technol. **12**, 309 (1975)

J. Berakdar, Phys. Rev. Lett. **83**(24), 5150 ± 5153 (1999)

J. Berakdar, S.N. Samarin, R. Herrmann, J. Kirschner, Phys. Rev. Lett. **81**, 3535 (1998)

J. Berakdar, M.P. Das, Phys. Rev. A **56**, 1403 (1997)

P. Blaha, K. Schwarz, G.K.H. Madsen, D. Kvasnicka, J. Luitz, WIEN2 k, (Karlheinz Schwarz, Techn. Universit˝at Wien, Austria) (2001). ISBN 3-9501031-1-2

C. Bowles, A.S. Kheifets, V.A. Sashin, M. Vos, E. Weigold, J. Electron Spectrosc. Relat. Phenom. **141**, 95–104 (2004)

A.M. Bradshow, D. Menzel, M. Steinkilberg, Jpn. J. Appl. Phys. Suppl. 2, Pt. 2, 841–846 (1974)

Y.Q. Cai, M. Vos, P. Storer, A.S. Kheifets, I.E. McCarthy, E. Weigold, Solid State Commun. **95**, 25 (1995)

W.A. Caliebe, J.A. Soininen, E.L. Shirley, C.-C. Kao, K. Hämäläinen, Phys. Rev. Lett. **84**, 3907 (2000)

R. Camilloni, A.G. Guidoni, R. Tiribelli, G. Stefani, Phys. Rev. Lett. **29**, 618 (1972)

A. D'Andrea, R. Del Sole, Surf. Sci. **71**, 306 (1978)

S.N. Davydov, M.M. Danilov, V.V. Korablev, J. Tech. Phys. (Russian) **69**(1), 109 (1999)

J.R. Dennison, A.L. Ritter, J. Electron Spec. Relat. Phenom. **77**, 99–142 (1996)

G. Di Filippo, D. Sbaraglia, A. Ruocco, G. Stefani, Phys. Rev. B **94**, 155422 (2016)

I. Egri, Solid State Commun. **44**, 563 (1982)

H. Ehrhardt, M. Schulz, T. Tekaat, K. Willmann, Phys. Rev. Lett. **22**, 89 (1969)

M. Erbudak, B. Reihl, Appl. Phys. Lett. **33**, 584 (1978)

D. Ewing, F. Seitz, Phys. Rev. **50**, 760 (1936)

R. Feder, H. Gollisch, Solid State Commun. **119**, 625 (2001)

B. Feuerbacher, B. Fitton, Phys. Rev. Lett. **29**(12), 786 (1972)

B. Feuerbacher, R.F. Willis, Phys. Rev. Lett. **37**(7), 446 (1976)

J.R. Fields, P.C. Gibbons, S.E. Schnatterly, Phys. Rev. Lett. **38**, 430 (1977)

A.E. Glassgold, G. Ialongo, Phys. Rev. **175**, 151 (1968)

F. Golek, E. Bauer, Surf. Sci. **369**, 415 (1996)

F. Golek, W.J. Sobolewski, Phys. Status Solidi (b) **216** R1 (1999)

F. Golek, Phys. Status Solidi (b) **177**, K5 (1993)

H. Gollisch, D. Meinert, X. Yi, R. Feder, Solid State Commun. **102**, 317 (1997)

H. Gollisch, R. Feder, J. Phys.: Condens. Matter **16**, 2207 (2004)

T.B. Grimley, Proc. Phys. Soc. Lond. **71**, 749 (1958)

G. Hansheng, H. Kawanowa, R. Souda, Appl. Surf. Sci. **158**, 163 (2000)

A.I. Gusarov, S.V. Murashov, Surf. Sci. **320**, 36 (1994)

M. Hattass, T. Jalowy, A. Czasch, Th Weber, T. Jahnke, S. Schössler, LPh Schmidt, O. Jagutzki, R. Dörner, H. Schmidt-Böcking, Rev. Sci. Instrum. **75**, 2373 (2004)

M. Hattass, T. Jahnke, S. Schössler, A. Czasch, M. Schöffler, LPhH Schmidt, B. Ulrich, O. Jagutzki, F.O. Schumann, C. Winkler, J. Kirschner, R. Dörner, H. Schmidt-Böcking, Phys. Rev. B **77**, 165432 (2008)

P. Hayes, M.A. Bennett, J. Flexman, J.F. Williams, Phys. Rev. B **38**, 13371 (1988)

O. Hemmers, S.B. Whitfield, P. Glans, H. Wang, D.W. Lindle, Rev. Sci. Instrum. **69**, 3809 (1998)

H. Burton, J. Liesegang, S.D. Smith, Phys. Rev. B **19**, 3004 (1979)

R. Herrmann, S. Samarin, H. Schwabe, J. Kirschner, Phys. Rev. Lett. **81**(10), 2148 (1998)

H. Chiuji, Progr. Theoret. Phys. **21**(1), 113 (1959)

S. Hüfner, *Photoelectron Spectroscopy* (Springer, Berlin, Heidelberg, 1996)

S. Iacobucci, L. Marassi, R. Camilloni, B. Marzilli, S. Nannarone, G. Stefani, J. Electron Spectrosc. Relat. Phenom. **76**, 109 (1995a)

S. Iacobucci, L. Marassi, R. Camilloni, S. Nannarone, G. Stefani, Phys. Rev. B **51**, 10252 (1995b)

O. Jagutzki, A. Cerezo, A. Czasch, R. Dörner, M. Hattass, M. Huang, V. Mergel, U. Spillmann, K. Ullmann-Pfleger, T. Weber, H. Schmidt-Böcking, G.D.W. Smith, IEEE Trans. Nucl. Sci. **49**, 2477 (2002)

K. Kimura, G. Andou, K. Nakajima, *Nuclear Instruments & Methods in Physics Research, Section Beam Interactions with Materials and Atoms*, vol. 164, 2000, p. 933

J. Kirschner, O.M. Artamonov, S.N. Samarin, Phys. Rev. Lett. **75**, 2424 (1995)

J. Kirschner, O. Artamonov, A. Terekhov, Phys. Rev. Lett. **69**, 1711 (1992)

J. Kirschner, G. Kerhervé, C. Winkler, Rev. Sci. Inst. **79**, 073302 (2008)

K.A. Kouzakov, J. Berakdar, Phys. Rev. A **85**, 022901 (2012)

U. Kolac, M. Donath, K. Ertl, H. Liebl, V. Dose, Rev. Sci. Instrum. **59**(9), 1933 (1988)

S.A. Komolov, *Total Current Spectroscopy* (Gordon and Breach Science Publishers, 1992)

D.A. Lapiano-Smith, E.A. Eklund, F.J. Himpsel, Appl. Phys. Lett. **59**, 2174 (1991)

V.G. Levin, V.G. Neudachin, Y.F. Smirnov, Phys. Status Solidi (b) **49**, 489 (1972)

E. Luscher Paul, F.M. Propst, J. Vac. Sci. Technol. **14**(1), 400 (1977)

Y.A. Mamaev, L.G. Gerchikov, Y.P. Yashin, D.A. Vasiliev, V.V. Kuzmichev, V.M. Ustinov, A.E. Zhukov, V.S. Mikhrin, A.P. Vasiliev, Appl. Phys. Letts. **93**, 081114 (2008)

I.E. McCarthy, E. Weigold, Phys. Rep. **27**, 277 (1976)

M. Michaels Amy, M. Nirmal, L.E. Brus, J. Am. Chem. Soc. **121**(43), 9932 (1999)

H. Morgner, Surf. Sci. **420**, 95 (1999)

A. Morozov, J. Berakdar, S.N. Samarin, F.U. Hillebrecht, J. Kirschner, Phys. Rev. B **65**, 104425 (2002)

D.R. Mullins, S.H. Overbury, Surf. Sci. **210**, 481 (1989)

V.G. Neudachin, G.A. Novoskol'tseva, Y.F. Smirnov, Sov. Phys. JETP **28**, 540 (1969)

J.B. Pendry, *Low Energy Electron Diffraction* (Academic Press, London, New York, 1974)

D.T. Pierce, R.J. Celotta, G.-C. Wang, W.N. Unertl, A. Galejs, C.E. Kuyatt, S.R. Mielczarek, Rev. Sci. Instrum. **51**, 478 (1980)

D.T. Pierce, F. Meier, Phys. Rev. B **13**, 5484 (1976)

D.T. Pierce, F. Meier, P. Zürcher, Appl. Phys. Lett. **26**, 670 (1975)

R.T. Poole, J.G. Jenkin, J. Liesegang, R.C.G. Leckey, Phys. Rev. B **11**, 5179 (1975)

B. Reihl, M. Erbudak, D.M. Campbell, Phys. Rev. B **19**, 9358 (1979)

M. Rohlfing, S.G. Louie, Phys. Rev. Lett. **81**, 2312 (1998)

G. Roy, G. Singh, T.E. Gallon, Surf. Sci. **152**(153), 1042 (1985)

S. Samarin, S. Herrmann, H. Schwabe, O. Artamonov, J. Electron Spectrosc. Relat. Phenom. **96**, 61 (1998)

S. Samarin, J. Berakdar, R. Herrmann, H. Schwabe, O. Artamonov, J. Kirschner, J. Phys. IV, Fr. **9**, Pr6–137 (1999)

S. Samarin, J. Berakdar, O. Artamonov, H. Schwabe, J. Kirschner, Surf. Sci. **470**, 141 (2000)

S.N. Samarin, O.M. Artamonov, H. Schwabe, J. Kirschner, in *The Book*: *Coincidence Studies of Electron and Photon Impact Ionization*, ed. by T.W. Colm, H.R.J. Walters (Plenum Publishing Corporation, 1997), p. 271

S. Samarin, J. Berakdar, A. Suvorova, O.M. Artamonov, D.K. Waterhouse, J. Kirschner, J.F. Williams, Surf. Sci. **548**, 187 (2004a)

S. Samarin, O.M. Artamonov, A.A. Suvorova, A.D. Sergeant, J.F. Williams, Solid State Commun. **129**, 389 (2004b)

S.N. Samarin, O.M. Artamonov, D.K. Waterhouse, J. Kirschner, A. Morozov, J.F. Williams, Rev. Sci. Instrum. **74**, 1274 (2003)

S. Samarin, O.M. Artamonov, P. Guagliardo, L. Pravica, A. Baraban, F.O. Schumann, J.F. Williams, J. Electron Spectrosc. Relat. Phenom. **198**, 26 (2015)

R. Schlaf, B.A. Parkinson, P.A. Lee, K.W. Nebesny, G. Jabbour, B. Kippelen, N. Peyghambarian, N.R. Armstrong, J. Appl. Phys. **84**, 6729 (1998)

F.O. Schumann, C. Winkler, J. Kirschner, F. Giebels, H. Gollisch, R. Feder, PRL **104**, 087602 (2010)

K.S. Shin, H.W. Kim, J.W. Chung, Surf. Sci. **385**, L978 (1997)

G. Singh, T.E. Gallon, Solid State Comm. **51**(5), 281 (1984)

Y.F. Smirnov, V.G. Neudachin, JETP Lett. **3**, 192 (1966)

T. Straub, R. Claessen, P. Steiner, S. Hüfner, V. Eyert, K. Friemelt, E. Bucher, Phys. Rev. B **55**(N20), 13473 (1997)

A.V. Subashievy, Y.A. Mamaev, Y.P. Yashin, J.E. Clendenin, SLAC-PUB-8035 (1998)

Y. Uehara, T. Ushiroku, S. Ushioda, Y. Murata, Jpn. J. Appl. Phys. Part 1 **29**, 2858 (1990)

M.A. Van Hove, W.H. Weinberd, C.-M. Chan, *Low Energy Electron Diraction*, Springer Series in Surface Science (Springer, Berlin, 1986)

W. Chao, Y. Kang, L. Weaver, Z. Chang, Rev. Sci. Instrum. **80**, 075101 (2009)

E. Weigold, S.T. Hood, P.J. Teubner, Phys. Rev. Lett. **30**, 475 (1973)

S.-L. Weng, E.W. Plummer, T. Gustafsson, Phys. Rev. B **18**, (4), 1718 (1978)

S.M.W. Werner, A. Ruocco, F. Offi, S. Iacobucci, W. Smekal, H. Winter, G. Stefani, Phys. Rev. B **78**, 233403 (2008)

S.M.W. Werner, F. Salvat-Pujol, W. Smekal, R. Khalid, F. Aumayr, H. Störi, A. Ruocco, G. Stefani, Appl. Phys. Letts. **99**, 184102 (2011)

J.F. Williams, J. Phys. B **8**, 2191 (1975)

Chapter 3
Spin-Polarized (e,2e) Spectroscopy of Surfaces

Abstract Two-electron coincidence spectroscopy with spin-polarized incident electrons is an efficient instrument for studying exchange and spin-orbit interaction on surfaces, thin films and multi-layered structures. This technique can visualize the "exchange-correlation hole" and probe quantum correlation of spin-entangled electron pairs scattered from a surface. Numerous examples of the application of spin-polarized (e,2e) spectroscopy for studying spin-dependent phenomena in magnetic and nonmagnetic surfaces show how this technique may characterize and develop spin-active nanostructures.

When spin-polarized electrons are used as the incident electron in the (e,2e) reaction on surfaces a new dimension is added to this process that highlights the role of electron spin in the electron-electron scattering on surfaces. Present experimental capability does not allow the measurements of the spin states of the two correlated electrons resulting from the (e,2e) scattering. However, the cross section of such reaction (intensity, or number of electron pairs per incident electron) may depend on the spin orientation of the incident beam. Such dependence can be measured and this is the objective here of the "spin-polarized (e,2e) spectroscopy of surfaces". Preliminary remarks indicate the basic description of spin-polarized electrons and role of spin-orbit interaction and exchange in electron scattering measurements. Detailed description of spin-polarized electrons and their scattering can be found in monographs (Kessler 1985; Feder 1981, 1985; Kirschner 1985a, b) and review papers (Hopster et al. 1990). Here the roles of spin-orbit and exchange effects in (e,2e) scattering from magnetic and nonmagnetic surfaces and associated experimental results are discussed.

© Springer Nature Switzerland AG 2018   87
S. Samarin et al., *Spin-Polarized Two-Electron Spectroscopy of Surfaces*, Springer
Series in Surface Sciences 67, https://doi.org/10.1007/978-3-030-00657-0_3

3.1 Description of Polarized Electrons and Spin-Dependent Interactions

In this section the general description of spin-polarized electron beam and spin-dependent electron scattering by Kessler (1985) is followed for completeness.

3.1.1 Spin Operator

The spin of an electron can be represented by the quantum mechanical operator:

$$\mathbf{s} = (s_x, s_y, s_z) \tag{3.1}$$

which must satisfy the cyclic commutation rule:

$$s_x s_y - s_y s_x = -i\hbar s_z \tag{3.2}$$

The spin operator can be represented as (3.3) through the use of the Pauli spin matrix $\boldsymbol{\sigma} = (\sigma_x, \sigma_y, \sigma_z)$; with components $\sigma_x = \begin{bmatrix} 0 & 1 \\ 1 & 0 \end{bmatrix}$, $\sigma_y = \begin{bmatrix} 0 & -i \\ i & 0 \end{bmatrix}$, $\sigma_z = \begin{bmatrix} 1 & 0 \\ 0 & -1 \end{bmatrix}$:

$$\mathbf{s} = \frac{\hbar}{2}\boldsymbol{\sigma} \tag{3.3}$$

The spin-state of an electron can be described in terms of the spin projection along a given axis, here the projection of spin along the z-axis is considered. Equation (3.4) shows the spin-function, χ, for the spin projected onto the z-axis with complex coefficients a_1 and a_2.

$$\chi = a_1|\alpha\rangle + a_2|\beta\rangle \tag{3.4}$$

Then using basis functions $|\alpha\rangle = \begin{pmatrix} 1 \\ 0 \end{pmatrix}$ and $|\beta\rangle = \begin{pmatrix} 0 \\ 1 \end{pmatrix}$ as eigenfunctions of σ_z shows

$$\chi = a_1\begin{pmatrix} 1 \\ 0 \end{pmatrix} + a_2\begin{pmatrix} 0 \\ 1 \end{pmatrix} = \begin{pmatrix} a_1 \\ a_2 \end{pmatrix} \tag{3.5}$$

The quantum mechanical probability of finding an electron in the "spin-up" state $+\frac{\hbar}{2}$ or the "spin-down" state $-\frac{\hbar}{2}$ with respect to the z-axis is given by $|a_1|^2$ and $|a_2|^2$ respectively. The operator s^2, however, which has the eigenvalue $s(s+1)\hbar^2 = 3\,h^2/4$, commutes with the components s_x, s_y, and s_z. Therefore its eigenvalue can be

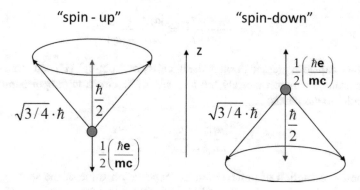

Fig. 3.1 Graphical representation of the statement "spin in Z-direction"

measured simultaneously with those of any of the components of **s**. Figure 3.1 shows a graphical representation of the relation between a spin vector of length $s = \hbar\sqrt{\frac{3}{4}}$ lying on a conical surface around the z-axis such that the projection of this vector onto the z-axis has the required value of $\pm\frac{\hbar}{2}$. The two other components are not known. However, it is known that

$$s_x^2 + s_y^2 + s_z^2 = \hbar^2 \frac{3}{4} \quad \text{(see Fig. 3.1)}.$$

For an electron system in a pure spin state (that is, all electrons are in the same spin state) the spin function χ corresponding to the state with the spin direction along the unit vector $\mathbf{e}(\theta, \varphi)$ (where θ is the polar angle with respect to Z-axis, and φ is the azimuthal angle with respect to X-axis) is determined by the eigenvalue equation ($\boldsymbol{\sigma} \cdot \mathbf{e}) \cdot \chi = \lambda \cdot \chi$. The solution of this equation is:

$$\chi = \begin{pmatrix} a_1 \\ a_2 \end{pmatrix} = \begin{pmatrix} \cos\frac{\theta}{2} \\ \sin\frac{\theta}{2} \cdot e^{i\phi} \end{pmatrix} \text{ for } \lambda = 1 \text{ and } \chi = \begin{pmatrix} a_1 \\ a_2 \end{pmatrix} = \begin{pmatrix} \sin\frac{\theta}{2} \\ -\cos\frac{\theta}{2} \cdot e^{i\phi} \end{pmatrix} \text{ for } \lambda = -1.$$

3.1.2 Polarization of an Electron Beam

The polarisation of a beam of electrons is defined by the expectation value of the spin operator:

$$\mathbf{P} = \frac{\langle \chi | \boldsymbol{\sigma} | \chi \rangle}{\langle \chi | \chi \rangle} \tag{3.6}$$

The projection of this vector onto a given axis provides the degree of polarisation. P_z, for example, is the degree of polarisation along the z-axis and is given by:

$$P_z = \frac{\langle\chi|\sigma_z|\chi\rangle}{\langle\chi|\chi\rangle} = \frac{|a_1|^2 - |a_2|^2}{|a_1|^2 + |a_2|^2} \tag{3.7}$$

Thus P_z can be expressed through the intensity of "spin-up" (I^+) and "spin-down" (I^-) electrons, or more precisely the intensity of electrons with spin parallel and antiparallel to the z-axis:

$$P_z = \frac{|a_1|^2 - |a_2|^2}{|a_1|^2 + |a_2|^2} = \frac{I^+ - I^-}{I^+ + I^-} \tag{3.8}$$

Since the definition of polarization as "expectation value of the spin operator" is not Lorentz invariant it is only valid in the rest frame of the electrons. However, relativistic corrections are suitably small so that at kinetic energies below E = 0.51 meV ($E \ll mc^2$) they can be neglected (Kessler 1985).

A "real world" beam of electrons will only be partially polarised. Under these conditions the beam can be considered as a mixture of n pure spin-states, such that (3.6) becomes:

$$\mathbf{P}^{(n)} = \frac{\langle\chi^{(n)}|\sigma|\chi^{(n)}\rangle}{\langle\chi^{(n)}|\chi^{(n)}\rangle} \tag{3.9}$$

thus, the total polarisation becomes the sum over all n states:

$$\mathbf{P} = \frac{\sum\limits_n \langle\chi^{(n)}|\sigma|\chi^{(n)}\rangle}{\sum\limits_n \langle\chi^{(n)}|\chi^{(n)}\rangle} \tag{3.10}$$

Introducing density matrix $\rho = \sum\limits_n |\chi^{(n)}\rangle\langle\chi^{(n)}|$ and using the matrix operator trace:

$$tr(\rho) = \sum\limits_n \langle\chi^{(n)}|\chi^{(n)}\rangle = \sum\limits_n \left(\left|a_1^{(n)}\right|^2 + \left|a_2^{(n)}\right|^2\right) \tag{3.11}$$

one can simplify the total polarisation to be:

$$\mathbf{P} = \frac{tr(\rho\sigma)}{tr(\rho)} \tag{3.12}$$

or for Cartesian components:

$$P_i = \frac{tr(\rho\sigma_i)}{tr(\rho)} \quad i = x, y, z \tag{3.13}$$

The explicit form for the spin-density matrix of the subsystem n $\rho^{(n)}$ is:

$$\rho^{(n)} = \begin{pmatrix} a_1^{(n)} \\ a_2^{(n)} \end{pmatrix} \cdot \begin{pmatrix} a_1^{(n)*} & a_2^{(n)*} \end{pmatrix} = \chi\chi^\dagger \tag{3.14}$$

and the total spin-density matrix is then given by:

$$\rho = \sum_n \rho^{(n)} = \sum_n \begin{pmatrix} \left|a_1^{(n)}\right|^2 & a_1^{(n)}a_2^{(n)*} \\ a_1^{(n)*}a_2^{(n)} & \left|a_2^{(n)}\right|^2 \end{pmatrix} \tag{3.15}$$

Using the derivation of (3.11) the total spin-density matrix can be expressed as:

$$\frac{\rho}{tr(\rho)} = \frac{1}{2}\begin{pmatrix} 1+P_z & P_x - iP_y \\ P_x + iP_y & 1 - P_z \end{pmatrix} = \frac{1}{2}(1+\mathbf{P}\boldsymbol{\sigma}) \tag{3.16}$$

The density matrix assumes its simplest (diagonal) form if the direction of the resultant polarization is taken as the Z axis of the coordinate system, then $P_x = P_y = 0$, $P = P_z$:

$$\text{and} \quad \rho = \frac{1}{2}\begin{pmatrix} 1+P & 0 \\ 0 & 1-P \end{pmatrix} \quad \text{(normalized spin functions are used)} \tag{3.17}$$

This form of the density matrix illustrates again the meaning of P:

$$P = \frac{N_\uparrow - N_\downarrow}{N_\uparrow + N_\downarrow}, \tag{3.18}$$

where $N_\uparrow$ and $N_\downarrow$ are the number of measurements that yield the value $+\hbar/2$ or $-\hbar/2$ for the spin projection. Indeed, since $\left|a_1^{(n)}\right|^2$ is the probability that the eigenvalue $+\hbar/2$ will be obtained from a spin measurement in the Z direction on the n-th subsystem, the probability is $\sum_n w^{(n)}\left|a_1^{(n)}\right|^2$ that this measurement on the total beam will give the value $+\hbar/2$. This probability can also be expressed as $N_\uparrow/(N_\uparrow + N_\downarrow)$. Then

$$\sum_n w^{(n)}\left|a_1^{(n)}\right|^2 = N_\uparrow/(N_\uparrow + N_\downarrow) = \frac{1}{2}(1+P) \quad \text{and} \tag{3.19}$$

$$\sum_n w^{(n)}\left|a_2^{(n)}\right|^2 = N_\downarrow/(N_\uparrow + N_\downarrow) = \frac{1}{2}(1-P). \tag{3.20}$$

It follows that:

$$P = \left(\frac{N_\uparrow - N_\downarrow}{N_\uparrow + N_\downarrow}\right). \tag{3.21}$$

3.1.3 Influence of Electron-Electron Scattering on the Polarization of an Electron Beam

Consider an idealized case of scattering of free primary electrons with polarization vector "up" from free target electrons with an arbitrarily given polarization with the z axis directed up. The polarization vector of the target electron is directed before scattering at the polar angle Ω relative to the z axis and at the azimuthal angle φ relative to the x axis in the xy plane. Assume that primary and valence electrons are totally polarized, i.e. the polarization vector length is equal to one and in the considered non-relativistic energy range the spin part of the wave function does not depend on the space coordinates (Mott and Massey 1965). Without loss of generality the spin part of the electron pair wave function $|\chi_{12}\rangle$ may be presented as a superposition in terms of four Bell two-particle states (Horodecki et al. 2009), one anti-symmetric singlet state and three symmetric triplet states. The full wave function $|\psi\rangle$ then consists either of a singlet spin component and symmetric space wave function ψ_s or of three triplet symmetric spin components and an anti-symmetric space wave function ψ_a. To determine the polarization of a state, for example, of the first electron of the pair, it is necessary to construct the density matrix for the system of two interacting particles $\rho = |\psi\rangle\langle\psi|$ and take the trace to obtain the one-particle reduced density matrix for the first particle $\rho_1 = Tr_2(\rho)$ (Blum 1996). The spin state of the particle is fully described by the polarization vector $\mathbf{P}$ (i.e. the expectation value of the Pauli spin operator), which is calculated as an expectation of the projectors on the corresponding coordinate axes $\mathbf{P} = \boldsymbol{\sigma}$, where σ_i are Pauli matrices and $P_i = \sigma_i = Tr_1(\sigma_i\rho_1)$, $i = (x, y, z)$. Consequently,

$$P_z = N^2\left[\cos^2\left(\frac{|\Omega|}{2}\right)|\psi_a|^2 + \frac{1}{2}\sin^2\left(\frac{|\Omega|}{2}\right)\left(\psi_a\ \psi_s^* + \psi_a^*\ \psi_s\right)\right], \qquad (3.22)$$

$$P_x = -N^2\frac{1}{4}\sin|\Omega|\left[-2\cos\varphi|\psi_a|^2 + e^{-i\varphi}\psi_a\ \psi_s^* + e^{i\varphi}\psi_a^*\ \psi_s\right], \qquad (3.23)$$

$$P_y = -i\,N^2\frac{1}{4}\sin|\Omega|\left[2i\,\sin\varphi|\psi_a|^2 + e^{-i\varphi}\psi_a\ \psi_s^* - e^{i\varphi}\psi_a^*\ \psi_s\right], \qquad (3.24)$$

where normalization factor $N = \left[\cos^2(|\Omega|/2)|\psi_a|^2 + 1/2\sin^2(|\Omega|/2)\left(|\psi_a|^2 + |\psi_s|^2\right)\right]^{-\frac{1}{2}}$.

The scattering amplitude in the centre of mass system when considering only the dependence on the scattering angle (θ): (Mott and Massey 1965) is given by:

$$f(\theta) \sim \csc[\theta/2]^2\exp\left[-i\alpha\log[1 - \cos[\theta]]\right], \qquad (3.25)$$

where $= \frac{1}{v_{rel}}$ is dimensionless factor, and v_{rel} is the relative electron velocity in atomic units and θ is the scattering angle in the system of centre of mass of the interacting electrons.

The anti-symmetric and symmetric wave functions are.

$$\psi_a[\theta] = f[\theta] - f[(\pi - \theta)],$$
$$\psi_s[\theta] = f[\theta] + f[(\pi - \theta)]. \tag{3.26}$$

If the valence electron mass is equal to the free electron mass, then $\theta = 2\theta_{\text{lab}}$. Detailed mathematical computations are given in (Artamonov et al. 2015).

The results of the modelling of the polarized electron scattering are presented in Fig. 3.2. The primary electrons polarized vertically along the z axis interact with the target electrons with the polarization vector directed at a polar angle Ω to the vertical axis, while the azimuthal angle $\varphi = 0°$. The interaction between the electrons is pure Coulombic and does not depend on the spin coordinates. The interaction intensity is described by the scattering angle θ_{lab} in the laboratory coordinate system or θ in the center of mass frame. If the valence electron mass is equal to the free electron mass then $\theta = 2\theta_{\text{lab}}$. The polarization vector components of the scattered electron are presented in Fig. 3.2 in parametric dependence on the scattering angle θ for two values of the angle Ω ($\Omega = 45°$—upper panels, $\Omega = 135°$ – lower panels). The relative velocity of the interacting electrons is $v_{\text{rel}} = 1.5$ in atomic velocity units (which corresponds to the relative kinetic energy of the interacting electrons ≈ 30 eV). The scattering angle θ changes from 0 to π in the centre-of-mass frame. The polarization vector of scattered electron depends on the scattering angle θ and the angle Ω between the directions of polarization vectors of the electrons before scattering. The choice of the azimuthal angle $\varphi = 0$ does not lead to loss of generality of results because of the azimuthal symmetry of the polarization vector. Dashed arrows in the upper-left panel indicate the spin directions of the primary and valence electrons respectively and the angle Ω is equal $\Omega = 45°$. The solid line is an envelope of all possible vector components of scattered electron.

Figure 3.2 (upper panels) represents vector components of scattered electron as a function of the scattering angle in the plane $P_x - P_z$ (left panel) and in the plane $P_y - P_z$ (right panel). Projection of the polarization vector on the plane $P_x - P_z$ and on the plane $P_y - P_z$ changes from the spin direction of the primary electron to the spin direction of the valence electron and passes through the coordinate origin at $\theta = \pi/2$. Dashed arrows in the lower-left panel indicate the spin directions of the primary and valence electrons respectively at the $\Omega = 135°$. The solid line is an envelope of all possible vector components of scattered electron. Figure 3.2 (lower panels) represents vector components of scattered electron as a function of the scattering angle in the plane $P_x - P_z$ (left panel) and in the plane $P_y - P_z$ (right panel). Projection of the polarization vector on the plane $P_x - P_z$ and on the plane $P_y - P_z$ changes from the spin direction of the primary electron to the spin direction of the valence electron and passes through the coordinate origin at $\theta = \pi/2$. Note, the polarization vector of the scattered electron does not necessary lie completely in the same plane as the polarization vectors of the electrons before scattering which is determined by the azimuth angle φ.

Equation (3.24) describes creation of the P_y component of the polarization vector of the scattered electrons. In the case of $\varphi = 0$ and using definitions of the wave functions (3.26) we get $P_y = -i N^2 \frac{1}{2} \sin|\Omega| [f(\theta)f(\pi - \theta) - f(\theta)f(\pi - \theta)]$, where N

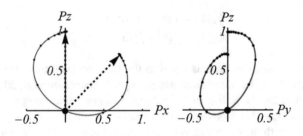

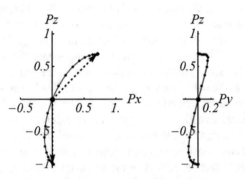

Fig. 3.2 Polarization vector $P(Px, Py, Pz)$ of the scattered electron in the parametric form as a function of the scattering angle θ for two values of the polar angle Ω between the polarization vectors of the interacting electrons before the scattering ($\Omega = 45°$—the upper panels, $\Omega = 135°$—the lower panels. The scattering angle changes in the range $0 < \theta < \pi$ with the step of 6° (see points on curves). Dashed arrows indicate the spin directions of the primary and valence electrons respectively. (Reprinted from Artamonov et al. (2015), Copyright (2015), with permission from Elsevier)

is the normalization factor, $f(\theta)$ is the direct scattering amplitude and $f(\pi - \theta)$ is a spin-flip (or exchange scattering) amplitude. The sign and value of P_y component of the polarization vector of the scattered electrons depends on the interference of direct and exchange amplitudes of electron-electron scattering. Both the modulus and the argument of the scattering amplitude depend on θ (see (3.25)). The dimensionless factor $\alpha = \frac{1}{v_{rel}}$ in the argument is the inverse relative electron velocity in atomic units and it decreases the interference of direct and exchange amplitudes with increasing of the relative velocity or energy of interaction. With the increase of the relative velocity of the interacting electrons $1/v_{rel} \ll 1$ the classic case is approached where the interference terms may be neglected. The polarization component P_y is negligibly small and all considered curves lie in one plane determined by the polarization vectors of the interacting electrons before the scattering at $\varphi = 0$.

The length of the vector defines the degree of polarization and it reduces to zero as the curve passes through the coordinate origin at the scattering angle $\theta = \pi/2$ at which point the scattered electrons are totally depolarized. Note also that the appearance of the polarization component P_y perpendicular to the $\varphi = 0$ plane is due to the interference of the scattering amplitudes and not to the electron polarization before the scattering. The space and spin coordinates in the discussed case are independent of each other. The electron-electron interaction is minimal at the scattering angles $\theta = 0$ and maximal at $\theta = \pi$, as easily seen in Fig. 3.2 where at $\theta = 0$ all three curves start at the point with the coordinates $P = (1,0,0)$, i.e. the polarization vector corresponds to the polarization of the primary beam. The polarization vector also passes through the coordinate origin, i.e. there exists a range of scattering parameters where the scattered electron beam is not polarized. It can be seen clearly in Fig. 3.2 that at the relatively small scattering angle $\theta \ll \pi/2$ the P_z polarization component is positive and coincides in sign with the corresponding component of the primary electrons. With the increase of the scattering angle $\theta \gg \pi/2$ the P_z polarization component changes sign and begins to approach to the initial polarization of the target electrons. Note that the polarization of the scattered electrons does not depend on the cross section of the electron-electron scattering and is determined by the symmetry of the two particle wave functions of scattering electrons. In the range of scattering angles $\theta \approx \pi/2$ the scattering is determined by the anti-symmetric spin wave function and is characterized by a high degree of entanglement of the spin states. In this way the depolarization of the scattered electrons at those particular scattering parameters is associated directly with the formation of entangled (non-separable) electron states. The destructive effect of entanglement on the individual quantum variables of a subsystem is of general nature and has its origin in the fact that the individual and collective observables do not commute.

3.2 Spin-Orbit Interaction in Electron Scattering

The spin-orbit effect is the direct result of the interaction between the spin and orbital momentum of an electron. Consider a simplified classical case of an incident electron with spin vector and magnetic moment scattered from a central potential. When this electron passes by the central potential it experiences a Coulomb attraction from a positive charge of the scattering centre which curves its path towards the centre. This motion generates another magnetic moment (orbital). The spin-orbit effect can be understood to the first order to occur as a result of the interaction between these two magnetic moments.

The Dirac equation is used to describe the behaviour of an electron in an electromagnetic field to include relativistic effects (Kessler 1985) as a product of two linear expressions considered separately so that:

$$\left(H^2 - c^2 \cdot \sum_{\mu} p_{\mu}^2 - m^2 c^2 \right) \cdot \Psi = 0, \tag{3.27}$$

(where $p_{\mu} = p_x, p_y, p_z$) can be transformed to

$$\left(H - c\sum_{\mu}\alpha_{\mu}p_{\mu} - \beta mc^2\right)\cdot\left(H + c\sum_{\mu}\alpha_{\mu}p_{\mu} + \beta mc^2\right)\Psi = 0, \qquad (3.28)$$

if the constant coefficients α_{μ} and β satisfy the relations:

$$\alpha_{\mu}\alpha_{\mu'} + \alpha_{\mu'}\alpha_{\mu} = 2\delta_{\mu\mu'} \qquad (3.29a)$$

$$\alpha_{\mu}\beta + \beta\alpha_{\mu} = 0 \qquad (3.29b)$$

$$\beta^2 = 1 \qquad (3.29c)$$

as seen by multiplication in (3.28).

If one can solve (3.30):

$$\left(H - c\sum_{\mu}\alpha_{\mu}p_{\mu} - \beta mc^2\right)\Psi = 0 \qquad (3.30)$$

(or the corresponding equation from (3.28) with the positive sign), then (3.28) is also solved. The linearized (3.30) has the advantage that it is of the first order in $\frac{\partial}{\partial t}$ like the Schrodinger equation. The derivatives with respect to the space coordinates are of the same order, which is necessary for relativistic covariance (Kessler 1985). The relations (3.29a, 3.29b, 3.29c) are solved using the following matrices, for example, which satisfy (3.29a, 3.29b, 3.29c):

$$\alpha_x = \begin{pmatrix} 0&0&0&1 \\ 0&0&1&0 \\ 0&1&0&0 \\ 1&0&0&0 \end{pmatrix}, \ \alpha_y = \begin{pmatrix} 0&0&0&-i \\ 0&0&i&0 \\ 0&-i&0&0 \\ i&0&0&0 \end{pmatrix}, \ \alpha_z = \begin{pmatrix} 0&0&1&0 \\ 0&0&0&-1 \\ 1&0&0&0 \\ 0&-1&0&0 \end{pmatrix}, \ \beta = \begin{pmatrix} 1&0&0&0 \\ 0&1&0&0 \\ 0&0&-1&0 \\ 0&0&0&-1 \end{pmatrix}$$

$$(3.31)$$

Thus, the Dirac equation for a free particle is, if one arbitrary chooses the left factor of (3.28):

$$\left[i\hbar\frac{\partial}{\partial t} + i\hbar c\left(\alpha_x\frac{\partial}{\partial x} + \alpha_y\frac{\partial}{\partial y} + \alpha_z\frac{\partial}{\partial z}\right) - \beta mc^2\right]\Psi = 0. \qquad (3.32)$$

Since α_{μ} and β are matrices (4×4), the (3.32) makes sense only if the wave function Ψ is the four-component vector:

$$\Psi = \begin{pmatrix} \Psi_1 \\ \Psi_2 \\ \Psi_3 \\ \Psi_4 \end{pmatrix} \qquad (3.33)$$

For an electron in external field (Φ, A) the Dirac equation takes the form:

$$\left[ih\frac{\partial}{\partial t} - e\Phi + (ihc\nabla + e\underline{A}) - \beta mc^2\right]\psi = 0,\tag{3.34}$$

where **p** and H in (3.27) are substituted by $\mathbf{p} + \frac{e}{c}\mathbf{A}$ and $H + e\Phi$.

Further insight into the standard approach is obtained by considering the non-relativistic limit with both kinetic and potential energies $\ll mc^2$. Under this condition the equation can be arranged back to nonlinear form into four key terms describing physical properties of the system and two components (3, 4) of four-component spinor can be neglected (Kirschner 1985a):

$$\left\{\underbrace{\frac{1}{2m}\left(\mathbf{P} - \frac{e}{c}\mathbf{A}\right)^2 + e\Phi}_{\text{I}} \quad \underbrace{-\frac{e\hbar}{2mc}\sigma\cdot curl\,\mathbf{A}}_{\text{II}} \quad \underbrace{+ i\frac{e\hbar}{4m^2c^2}\mathbf{E}\cdot\mathbf{p}}_{\text{III}} \quad \underbrace{-\frac{e\hbar}{4m^2c^2}\sigma[\mathbf{E}\times\mathbf{p}]}_{\text{IV}}\right\}\Psi = E\Psi$$

I	II	III	IV
Hamilton operator of non-relativistic Schrödinger eq. for spin-less particle in an electromagnetic field	Interaction of spin with magnetic field	Relativistic correction to the energy	General spin-orbit interaction

$$\tag{3.35}$$

Term IV is of primary concern in this section as it describes the spin-orbit interaction. This term can be simplified by taking a spherical coordinate system and consider a central potential $\Phi = \Phi(\mathbf{r})$, with electric field given by:

$$\mathbf{E} = -\left(\frac{1}{e}\right)\frac{d\Phi}{dr}\cdot\frac{\mathbf{r}}{r}\tag{3.36}$$

One can use (3.36) and the orbital angular momentum, $l = r \times p$, to determine the spin-orbit interaction energy, U_{so}, from term IV of (3.35):

$$U_{SO} = -\frac{e\hbar}{4m^2c^2}\sigma\cdot[\mathbf{E}\times\mathbf{p}] = \frac{1}{2m^2c^2}\frac{1}{r}\frac{d\Phi}{dr}(\mathbf{s}\,\mathbf{l})\tag{3.37}$$

If one takes the Coulomb potential of an electron interacting with a nucleus of charge Z, $\Phi(r) = -\frac{Ze^2}{r}$, the spin-orbit interaction energy can be approximated as:

$$U_{SO} \approx \frac{Z}{r^3}(\mathbf{s}\,\mathbf{l})\tag{3.38}$$

Thus the spin-orbit effect is proportional to the atomic number of the nucleus and the scalar product of spin and orbital angular momentum, as expected in the classical approximation due to the Coulomb interaction between the electron and the nucleus. This Z relationship for the spin-orbit effect has been repeatedly confirmed with heavy atoms (such as tungsten, gold) showing high spin-orbit interaction and with light atoms (such as iron and cobalt) showing weak spin-orbit interaction.

Fig. 3.3 Modified Coulomb scattering potential for "spin-up" ↑ and "spin-down" ↓ electrons passing the scattering centre on the left or on the right. In this case the spin is normal to the scattering plane with the solid "no spin" curve in provided for comparison (Kessler 1985)

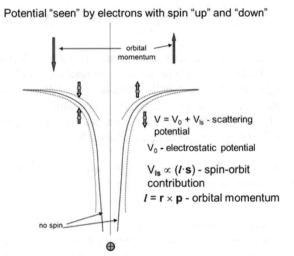

Potential "seen" by electrons with spin "up" and "down"

$V = V_0 + V_{ls}$ - scattering potential

V_0 - electrostatic potential

$V_{ls} \propto (\boldsymbol{l}\cdot\mathbf{s})$ - spin-orbit contribution

$\boldsymbol{l} = \mathbf{r} \times \mathbf{p}$ - orbital momentum

Scattering center

The spin-orbit interaction acts to modify the Coulomb potential to increase or decrease depending on the spin orientation of the incident electron. Classically this can be understood by the different direction of curvature as the electron passes either side of the ion core.

A modification of the "no spin" Coulomb scattering potential is shown graphically on Fig. 3.3. An example of the spin-orbit effect in the electron scattering from a surface is shown in Fig. 3.4, where SPEELS spectra measured in the left and right detectors are shown for spin-up and spin-down primary electrons. A spin-polarized electron beam with electron energy 65 eV and normal incidence is scattered from W(110) surface. Two diffracted beams (01) and (0–1) and secondary electrons are detected to indicate why the elastic maximum is dominant in the time-of-flight spectra of Fig. 3.4. The amplitude of the elastic maximum (seen as sharp peaks on all four spectra) depends on the spin orientation of the incident beam and it depends also on whether the left or right detector receives the electrons. Arrows next to the elastic maxima of the spectra shows polarization of the incident beam ("up" or "down"). It is clear from the Fig. 3.4 that for the same polarization of the incident beam the intensities of elastic maxima are different in the two detectors. For example, for spin-up incident beam polarization the intensity of elastic maximum is higher in the left detector than in the right one. For spin-down polarization of the incident electrons, the intensity of the elastic maximum in the right detector is higher than in the left one. The asymmetry (normalised difference) for each spectrum can be calculated as:

$$A_{so}(\theta) = \frac{I^+(\theta) - I^-(\theta)}{I^+(\theta) + I^-(\theta)} \qquad (3.39)$$

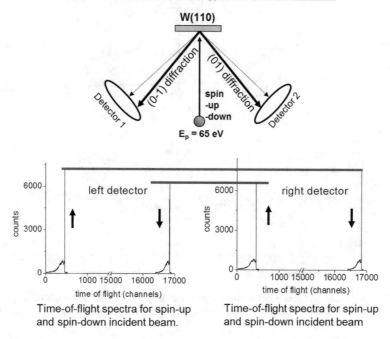

Fig. 3.4 Spin-polarized Low Energy Electron Diffraction from W(110) and SPEELS

The measured asymmetries are inversely symmetric at complementary angles, i.e. $A_{so}(\theta) = -A_{so}(-\theta)$. This left-right symmetry in scattering is expected after considering the nature of the spin-orbit effect and Fig. 3.3.

Thus, the asymmetry of elastic scattering shows left-right symmetry (while flipping the polarization of incident electrons) that is typical for the spin-orbit effect. Figure 3.5 shows the intensity asymmetries of the secondary emission spectra of W(110) recorded by the left (Det. 1) and right (Det. 2) detectors at normal incidence of spin-polarized electrons. Again, if the detectors exchange their positions (the sample is assumed to be mirror symmetric with respect to the plane perpendicular to the surface and to the scattering plane), the corresponding asymmetries change their signs, as it should be for spin-orbit asymmetry.

3.2.1 *Electron Scattering Cross Section in a Relativistic Case (Taking into Account Spin-Orbit Interaction)*

In analogy to nonrelativistic scattering theory, solutions are sought for the Dirac equation with the asymptotic form:

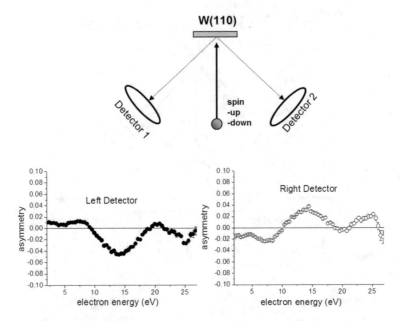

Fig. 3.5 Asymmetry of secondary emission spectra of W(110) recorded at complimentary angles (±50°) and normal incidence. The corresponding values of the asymmetries have opposite signs

$$\psi_\lambda \xrightarrow[r \to \infty]{} a_\lambda e^{ikz} + a'_\lambda(\theta, \phi)\frac{e^{ikr}}{r} \tag{3.40}$$

for the four components of the wave function ($\lambda = 1, 2, 3, 4$). From the general definition of the differential cross section and from the fact that the current density can be written as $\rho \cdot \mathbf{v} = |\psi\rangle\langle\psi|\mathbf{v}$ it follows:

$$\frac{d\sigma}{d\Omega}(\theta, \phi) \equiv \sigma(\theta, \phi) = \frac{\sum_{\lambda=1}^{4} |a'_\lambda(\theta, \phi)|^2}{\sum_{\lambda=1}^{4} |a_\lambda|^2} \tag{3.41}$$

Considering the scattering of electron waves with spin orientation along +Z or −Z directions and assuming that two small components of the wave function (ψ_3 and ψ_4) yield no additional information for the asymptotic solution of the scattering problem, it follows that:

$$\begin{pmatrix} \psi_1 \\ \psi_2 \end{pmatrix} \xrightarrow[r \to \infty]{} \begin{pmatrix} 1 \\ 0 \end{pmatrix} e^{ikz} + \begin{pmatrix} S_{11}(\theta, \varphi) \\ S_{21}(\theta, \varphi) \end{pmatrix} \frac{e^{ikr}}{r} \tag{3.42}$$

This expression takes account of the fact that the second component of the wave function is no longer necessarily zero after scattering, since the spin may change its direction due to spin-orbit coupling as described by the scattering amplitude S_{21}. The electron "sees" in its rest frame the moving charge of the scattering centre which means that it sees a current and thus a magnetic field, which acts upon its magnetic moment and may change its spin direction. This possibility is taken into account through the inclusion of, S_{21} which is therefore called the "spin-flip amplitude". For the incident wave (electron) with the other spin direction, one can expect a solution with the following asymptotic form:

$$\begin{pmatrix} \psi_1 \\ \psi_2 \end{pmatrix} \xrightarrow[r\to\infty]{} \begin{pmatrix} 0 \\ 1 \end{pmatrix} e^{ikz} + \begin{pmatrix} S_{12}(\theta, \varphi) \\ S_{22}(\theta, \varphi) \end{pmatrix} \frac{e^{ikr}}{r} \tag{3.43}$$

Detailed calculations are given by Kessler (1985) to show the solutions for scattered waves that satisfy (3.42) and (3.43). For the incident electron beam with spin along $+Z$ the scattering amplitudes are:

$$S_{11}(\theta, \varphi) = \frac{1}{2ik} \sum_{l=0}^{\infty} \left[(l+1)(e^{2i\eta_l} - 1) + l(e^{2i\eta_{-l-1}} - 1) \right] \cdot P_l(\cos\theta) \equiv f(\theta) \tag{3.44}$$

$$S_{21}(\theta, \varphi) = \frac{1}{2ik} \sum_{l=1}^{\infty} \left[(-e^{2i\eta_l} + e^{2i\eta_{-l-1}}) \right] \cdot P_l^1(\cos\theta)e^{i\phi} \equiv g(\theta) \cdot e^{i\phi} \tag{3.45}$$

For the incident electron beam with spin along—Z direction the scattering amplitudes S_{12} and S_{22} are:

$$S_{22} = S_{11} = f(\theta) \text{ and}$$
$$S_{12} = -S_{21}e^{-2i\varphi} = -g(\theta)e^{-i\varphi}.$$

Now, for the incident electron beam with an arbitrary spin direction:

$$A\begin{pmatrix} 1 \\ 0 \end{pmatrix} \cdot e^{ikz} + B\begin{pmatrix} 0 \\ 1 \end{pmatrix} \cdot e^{ikz} = \begin{pmatrix} A \\ B \end{pmatrix} \cdot e^{ikz}, \tag{3.46}$$

where A and B specify the spin direction in the rest frame. Then for the asymptotic form of the scattered wave one has:

$$A \cdot \begin{pmatrix} S_{11} \\ S_{21} \end{pmatrix} \cdot \frac{e^{ikr}}{r} + B\begin{pmatrix} S_{12} \\ S_{22} \end{pmatrix} \cdot \frac{e^{ikr}}{r} = \begin{pmatrix} Af - Bge^{-i\varphi} \\ Bf + Age^{i\varphi} \end{pmatrix} \cdot \frac{e^{ikr}}{r} = \begin{pmatrix} a_1' \\ a_2' \end{pmatrix} \cdot \frac{e^{ikr}}{r}. \tag{3.47}$$

Thus, from (3.41) the differential cross section is:

$$\sigma(\theta, \varphi) = \frac{|a_1'|^2 + |a_2'|^2}{|A|^2 + |B|^2} = |f|^2 + |g|^2 + \frac{-AB^*e^{i\varphi} + A^*Be^{-i\varphi}}{|A|^2 + |B|^2}(fg^* - f^*g), \tag{3.48}$$

which shows that for a polarized primary beam the scattering intensity generally depends on φ. The Sherman function is introduced as

$$S(\theta) = i\frac{fg^* - f^*g}{|f|^2 + |g|^2} \qquad (3.49)$$

and the cross section as follows:

$$\sigma(\theta, \varphi) = \left(|f|^2 + |g|^2\right) \cdot \left[1 + S(\theta)\frac{-AB^*e^{i\varphi} + A^*Be^{-i\varphi}}{i\left(|A|^2 + |B|^2\right)}\right] \qquad (3.50)$$

Apart from depending on the scattering angle θ, the amplitudes f and g depend also on the energy of the incident electrons and essentially on the scattering potential.

The spin-orbit asymmetry can be expressed using the two scattering amplitudes that were introduced above:

$$A_{so}(\theta) = i\frac{fg^* - f^*g}{|f|^2 + |g|^2} \qquad (3.51)$$

Thus, the spin-orbit asymmetry can be considered as the result of interference between the direct and spin-flip scattering amplitudes.

It is noted that the spin-orbit interaction not only causes an asymmetry in the scattering intensity of polarised electrons but can also cause polarisation in a beam of unpolarised electrons during scattering. In such a case the polarisation vector of the scattered electrons is normal to the scattering plane. The degree of polarisation, P', after scattering is identical to the intensity asymmetry, A_{so}, that would occur when scattering polarised electrons with polarisation normal to the scattering plane, i.e. the reverse of the polarised incident electron scattering experiment under spin-orbit coupling.

3.2.2 Spin-Orbit Effects in the (e,2e) Scattering on Surfaces

Since spin-orbit interaction (SOI) is the interaction of the spin of an electron with its orbital momentum it is essentially a single-electron effect in contrast to the exchange effect, where an ensemble (at least two) of *spin-half* particles experience the requirement of the Pauli Exclusion Principle, which results in so called "exchange interaction". Therefore, each of the four single electron states involved in the (e,2e) process: (i) incident electron (E_1, $\mathbf{k}_1$), valence electron (E_2, $\mathbf{k}_2$), and two outgoing electrons (E_3, $\mathbf{k}_3$) and (E_4, $\mathbf{k}_4$) may experience the SOI and make a contribution to the overall spin-orbit effect of the (e,2e) reaction. On the other hand, the SOI in a heavy-Z material results in the spin-splitting of electron energy bands and leads to a separation in momentum space of spin-up and spin-down states. It means that if the experimental conditions enable the incident spin-polarized electron to interact predominantly with spin-up or spin-down valence electrons, then the spin effect of such an (e,2e) reaction on non-magnetic surface would result from the exchange scattering. The spin-orbit effects in the (e,2e) reaction with spin-polarized incident electrons will be demonstrated using results of spin-polarized (e,2e) scattering from W(100), W(110) and Au(111) films on W(110).

Brief Theoretical Consideration

Extensive theoretical work on spin-polarized low-energy (e,2e) spectroscopy of non-magnetic surfaces (Gollisch et al. 1999, 2006; Giebels et al. 2013; Feder and Gollisch 2003 Photoem Rel Meth) predicted an observable spin-asymmetry in the (e,2e) cross section from tungsten. Outlined here is the method of calculation of the (e,2e) reaction cross section for spin-polarized low-energy electrons impinging on a non-magnetic crystalline surface (W(001)) and derived is the appropriate spin-dependent expressions (Gollisch et al. 1999).

The model of the (e,2e) reaction is constructed as follows. A primary electron is characterized at the source of electron beam by kinetic energy E_1, momentum k_1, and spin alignment $\sigma_1 = \pm$ with respect to some given axis e. This electron collides with a valence electron (with $E_2 < E_F$, where E_F is the Fermi energy). As a result of the collision, two directly produced outgoing electrons with kinetic energies E_3 and E_4 and momenta k_3 and k_4 activate the detectors. The energy E_2 of the valence electron is determined from the energies of the primary and the detected electrons by the conservation law: $E_1 + E_2 = E_3 + E_4$. The surface-parallel momentum $\mathbf{k}_i^{\parallel}$ with $i = 1; 2; 3; 4$ is a good quantum number for each of the four electron states because of the lattice periodicity parallel to the surface; $\mathbf{k}_1^{\parallel}$ and $\mathbf{k}_{3,4}^{\parallel}$ are the surface projections of the incident and the detected electrons momenta, and $\mathbf{k}_2^{\parallel}$ is determined by the conservation requirement that: $\mathbf{k}_1^{\parallel} + \mathbf{k}_2^{\parallel} = \mathbf{k}_3^{\parallel} + \mathbf{k}_4^{\parallel}$.

The collision of the incident electron and a valence electron in a W(001) surface is treated using a distorted-wave Born approximation formalism, with exchange (Gollisch et al. 1999). Four relevant quasi-one-electron states are determined as solutions to the Dirac equation using one-electron potential describing the interaction with the atomic nuclei and the ground-state electrons of the surface system. These wave functions cover the incident electron state (kinetic energy E_1, momentum $\mathbf{k}_1$, spin-state $\sigma_1 = \pm$), the valence electron (E_2, $\mathbf{k}_2$, σ_2) and the two outgoing electron states (E_3, $\mathbf{k}_3$, σ_3) and (E_4, $\mathbf{k}_4$, σ_4). For each of the four states there are two independent spinor solutions $\psi_i^{\sigma_i}(x)$, which are characterized by the spin label $\sigma_i = \pm$. For the primary electron state, the spin-quantization axis and the spin label σ_1 are fixed by the boundary condition at the spin-polarized electron source.

The primary electron state $\psi_1^{\sigma_1}(x)$ which is the usual relativistic LEED state, and $\psi_3^{\sigma_3}(x)$ and $\psi_4^{\sigma_4}(x)$, which are time-reversed LEED states, are calculated by means of a relativistic layer KKR method (for details, see e.g. chapter 4.3.5 of (Feder 1985)) with the appropriate boundary conditions. The valence electron states are obtained by matching—at the surface -linear combinations of bulk Bloch waves inside the crystal with exponentially decaying plane waves in the vacuum half-space. There are usually

$2N$ independent solutions $\psi_2^{n\sigma_2}(x)$ where the index $n=1, ..., N$ corresponds to the N outward-propagating bulk Bloch-wave pairs.

In first-order perturbation theory the reaction cross section is approximated in the form $|\langle 3, 4|H_{ee}|1, 2\rangle|^2$, where H_{ee} is the electron–electron interaction Hamiltonian, and $|3, 4\rangle$ and $|1, 2\rangle$ are antisymmetrized products of the above one-electron states.

For primary electrons with spin orientation σ_1 relative to an axis e (i.e. spin-polarization vector $P_1 = \sigma_1 \cdot e$) and fixed energy and momentum, and for spin-unresolved detection of the outgoing electrons in fixed directions, this leads to the following expression for the (e,2e) cross section ('intensity'):

$$I^{\sigma_1}(E_3, E_4) = \left(\frac{k_3 k_4}{k_1}\right) \sum_{\sigma_3 \sigma_4} \sum_{E_2, k_2^\parallel, n\sigma_2} \left|f_{1,2,3,4}^{\sigma_1, n\sigma_2, \sigma_3, \sigma_4} - g_{1,2,3,4}^{\sigma_1, n\sigma_2, \sigma_3, \sigma_4}\right|^2 \delta(E_1 + E_2 - E_3 - E_4)$$

$$\times \delta\left(k_1^\parallel + k_2^\parallel - k_3^\parallel - k_4^\parallel\right)$$

(3.52)

where f and g are direct and exchange ("spin-flip") scattering amplitudes:

$$f_{1,2,3,4}^{\sigma_1, n\sigma_2, \sigma_3, \sigma_4} = \int \psi_3^{\sigma_3 *}(x)\psi_4^{\sigma_4 *}(x')V(x, x')\psi_1^{\sigma_1}(x)\psi_2^{n\sigma_2}(x')d^3x d^3x'. \quad (3.53)$$

The expression for g is the same except that x and x' are interchanged in the first product term. The state subscripts i implicitly contain the quantum numbers E_i and $k_i^\parallel$. Since the spin of the outgoing electrons is not resolved, the observed cross section is the σ_3, σ_4 sum over cross sections involving states $\psi_3^{\sigma_3}$ and $\psi_4^{\sigma_4}$. Each of these partial cross sections consists of a sum over the independent valence states. $V(x,x')$ in (3.53) approximates the electron–electron interaction H_{ee} by the Coulomb interaction statically screened by the ground-state electrons of the target:

$$V(x, x') = \int d^3x'' \varepsilon^{-1}(x, x'')/|x'' - x'|, \quad (3.54)$$

where $\varepsilon(x, x'')$ is the dielectric function of the crystalline surface.

The asymmetry A of the cross section, upon reversal of the primary electron spin, is defined from the intensities $I^\pm$ in (3.52) as:

$$A(E_3, E_4) = \frac{(I^+ - I^-)}{(I^+ + I^-)}. \quad (3.55)$$

If, for a non-magnetic target system, spin - orbit coupling is absent in all four states, the spinor wave functions $\psi_i^{\sigma_i}(x)$ reduce to products of a scalar spatial wave function with a basic spinor, and one easily obtains from (3.52), (3.53) that $I^+ = I^-$ and therefore $A = 0$. SOC—in at least one of the participating one-electron states—is thus essential for the (e,2e) cross section to depend on the primary electron spin.

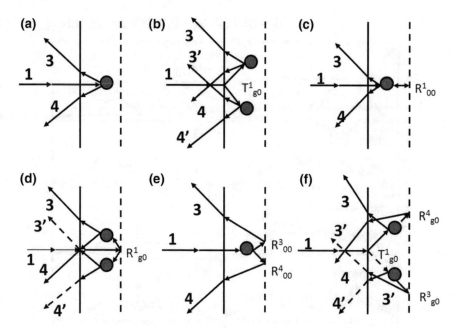

Fig. 3.6 Symbolic diagrams of typical scattering paths in the (e, 2e) process for normal incidence of the primary electron (labelled 1) and polar angles of 40° of the detected electrons (3 and 4). The vertical thin solid line indicates the surface. The filled circle symbolizes the collision with the valence electron. The vertical thin dashed lines stand for the atomic planes, and the R^i_{gg}, on their right-hand sides are the elastic reflection matrix elements, which are relevant in the individual diagrams. In diagrams (**d**) and (**f**), the thick solid and dashed lines depict two different paths, which are related to each other by symmetry. The refraction angle of electron 3 at the surface has been drawn equal to that of electron 4, which is actually the case only for $E_3 = E_4$. For $E_3 \neq E_4$, the diagrams are topologically equivalent to the ones shown, but due to the energy dependence of refraction at the surface—the internal angles of electron 3 are different from those of electron 4. (From Gollisch et al. (1999) 9555. © IOP Publishing. Reproduced with permission. All rights reserved)

    The theory takes into account the importance of multiple elastic scattering from the ion cores in both the primary and ejected electron states (Fig. 3.6). To explore the effects of these scattering events on each of the four electron states the authors artificially modified the wave functions such that they were able to selectively switch off the elastic scattering amplitudes (Fig. 3.7). They found that certain features require mostly elastic reflections in the primary electron state, while others require reflection in one of the ejected electron states. It is shown in the right column of Fig. 3.7.

    By taking into account the spin-state of each one-electron wave function the authors (Gollisch et al.1999) were able to selectively switch off the elastic scattering amplitudes and also control the presence of the spin-orbit coupling term in each of the one-electron states. This approach allowed the authors to construct energy sharing curves in which these parameters can be controlled to determine the origin of various scattering features. The results of these calculations are presented in Fig. 3.7.

$$E_1 = 24.6 \, \text{eV} \quad E_3 + E_4 = 18.5 \, \text{eV} \quad (E_F\text{-}1.5\text{eV})$$

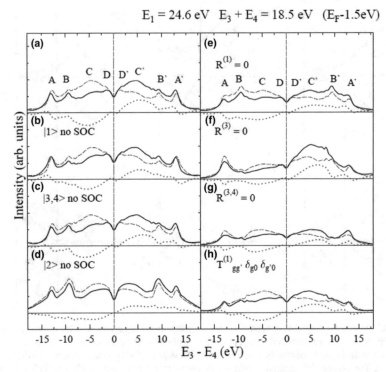

Fig. 3.7 Energy sharing curves of W(001) for I^+ (solid) and I^- (dashed), and the difference (dotted). Primary energy $E_1 = 24.6 \, \text{eV}$, with the total energy slice ranging from E_F to $E_F - 1.5 \, \text{eV}$. Comparison between the results of a 'complete calculation' and those in which the spin-orbit coupling and elastic modes are switched off. Note the numbering corresponds to 1: primary electron, 2: valence electron, 3 and 4: outgoing electrons. (From Gollisch et al. (1999) 9555. © IOP Publishing. Reproduced with permission. All rights reserved)

It is clear that the I^+ and I^- of the 'complete calculation' (Fig. 3.7a) change very little if the spin-orbit coupling is switched off for the incident electron (b) or the ejected electron (c). However, there is a significant effect when the spin-orbit coupling in the valence electron is removed (d). The authors next used typical (e,2e) scattering paths in order to identify which paths are responsible for the individual scattering features. From these results they determined that in the majority of peak asymmetries the spin-orbit coupling of the valence state is partly or fully responsible for the structure. This approach requires that upon meeting the surface the valence electron wave is reflected into Bloch waves moving into the crystal. However, due to spin-orbit coupling the reflected amplitudes of the Bloch waves will be different for different spin-states. This process can be viewed as slightly similar to spin-polarised LEED from the inside of the surface. As a result, the scattering amplitudes, from (3.52), $f^{\sigma_1, n\sigma_2, \sigma_3, \sigma_4}$ will be different for the two spin-states of the valence electron (σ_2^+ and σ_2^-), the exchange scattering amplitudes will also be different for each spin-state (Gollisch et al. 1999). The σ_1-dependency is maintained through the σ_2-summation in (3.40)

Fig. 3.8 Geometry of the (e,2e) scattering on W surface

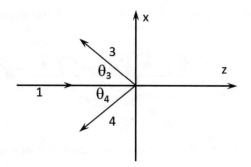

ensuring that the summation in scattering amplitude over the valence electron spin-state does not "smear out" this difference which shows up as a scattering intensity difference for different incident electron spins, i.e. I^+ and I^-. Thus, this theoretical approach predicts that a spin-asymmetry should be observed from tungsten using low-energy spin-polarised (e,2e) spectroscopy (Gollisch et al. 1999). They determined that, in the case of W(001), the spin-asymmetry in the spin-polarised (e,2e) scattering is due principally to the spin-orbit coupling in the valence electron state (d states). The latter appears plausible from the well-known substantial influence of spin-orbit coupling (SOC) on the valence band structure of W (Bylander and Kleinman 1984; Christensen and Feuerbacher 1974). The comparatively smaller influence of SOC in the excited states can be understood in the following way. Whilst the valence states have predominant d character ($l = 2$, the excited-state bands which can couple to the vacuum to form the LEED states are mostly sp-like ($l = 0; 1$) [cf. e.g. (Courths et al. 1999)], and SOC is overall stronger for the occupied valence d states than for the unoccupied sp states.

The contour plots of Fig. 3.7 exhibit an interesting symmetry property. Reflection at the $E_3 = E_4$ diagonal, i.e. interchanging E_3 and E_4, transforms I^+ into I^- and hence A into $-A$. This is readily understood from the geometry shown in Fig. 3.8. Reflection at the y, z—plane reverses the spin of the primary electron and interchanges the outgoing electrons 3 and 4, which implies that $I^+(E_3, E_4) = I^-(E_4, E_3)$.

3.3 Role of Electron Exchange in the Electron Scattering (Basics)

The Pauli Exclusion Principle states that the wave function for two spin ½ particles must be antisymmetric when the particles are exchanged. Thus, only two electrons of opposite spin quantum number can occupy any particular quantum state. In the low-energy regime this interaction can be treat as non-relativistic. Therefore, unlike the spin-orbit case, we can describe the exchange interaction with the Schrödinger equation.

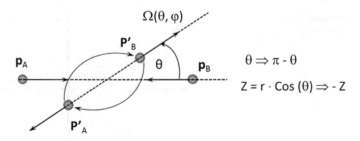

Fig. 3.9 Singlet exchange of two electrons in the centre-of-mass frame

Consider the interaction of two free electrons scattering in a centre-of-mass frame (see Fig. 3.9). This approach is equivalent to scattering from a central potential and one can use the ansatz employed in non-relativistic scattering theory to describe the scattered wave function of one electron from another (Landau and Lifshitz 1991).

$$\Psi \approx e^{ikz} + f(\theta)\frac{e^{ikr}}{r} \tag{3.56}$$

If spin-flip processes are excluded, as for pure exchange interactions (spin-orbit interaction is negligible), the spins are conserved, the wavefunction of the two-particle system can be factorised into spatial and spin components. The total spin of the system will be either 0 (singlet) for opposite spin electrons, or 1 (triplet) for parallel spin electrons. For the singlet state the spin component of the wave function is antisymmetric, thus the spatial component of the total wavefunction must be symmetric. This exchange in the centre of mass frame is shown graphically on Fig. 3.9.

The symmetrised wavefunction for the singlet state can be expressed as:

$$\Psi_s = V\left\{ e^{ikz} + e^{-ikz} + \left[f_s(\theta) + f_s(\pi - \theta)\right]\frac{e^{ikr}}{r}\right\}, \tag{3.57}$$

where V is a normalisation factor and the symmetrised scattering amplitude is given by $\left[f_s(\theta) + f_s(\pi - \theta)\right]$. In fact, this scattering amplitude can be presented as two components:

$$f(\theta) \equiv f_s(\theta) \qquad \text{direct scattering amplitude}$$
$$g(\theta) \equiv f_s(\pi - \theta) \text{ exchange scattering amplitude} \tag{3.58}$$

Note here that the "exchange scattering amplitude" $g(\theta)$ has nothing to do with the "spin-flip" amplitude $g(\theta)$ introduced in the section describing spin-orbit interaction.

Thus, the cross-section for the singlet state is simply:

$$\sigma_s = |f(\theta) + g(\theta)|^2 \tag{3.59}$$

In a triplet state the spin component of the total wave function is symmetric, thus the spatial component must be antisymmetric. Using the same methodology as for the singlet state we obtain the following wavefunction:

$$\Psi_t = V \left\{ e^{ikz} + e^{-ikz} + \left[f_t(\theta) - f_t(\pi - \theta) \right] \frac{e^{ikr}}{r} \right\}$$ (3.60)

with a cross-section equal to:

$$\sigma_t = |f(\theta) - g(\theta)|^2$$ (3.61)

Using the cross-sections, the exchange asymmetry can be determined as the normalised difference between the singlet and triplet cross-sections:

$$A_{ex} = \frac{\sigma_t - \sigma_s}{\sigma_t + \sigma_s}.$$ (3.62)

It should be noted that unlike the "left-right" symmetry of the spin-orbit interaction, $A_{so}(\theta) = -A_{so}(-\theta)$, the sign of the exchange asymmetry does not change for complimentary angles, i.e. $A_{ex}(\theta) = A_{ex}(-\theta)$ (see Fig. 3.10).

Thus, there is no left-right asymmetry for the exchange effect, only an "up-down" spin-asymmetry. Figure 3.10 shows asymmetries of SPEELS for 3 ML cobalt film on W(110), recorded by the left and right detectors at the sample magnetization "up" (M1, first row in Fig. 3.10) and "down" (M2, second row). Figure 3.10 represents Stoner excitation asymmetry, which is, by nature, due to electron exchange. In fact, the exchange interaction only depends on the relative spin orientation of the two electrons involved. That is, it depends on the mutual orientation of the polarization vector of the incident beam and the magnetic moment of the sample. Moreover, as the exchange interaction is a function of wave function symmetry it does not depend on the absolute spin orientations of the electrons with respect to the scattering plane, only on the relative spin orientations of the interacting electrons. The reader should note that the exchange effect is present in all electron/atom and electron/surface interactions, however the effect is most pronounced in ferromagnetic materials due to the imbalance of spin-up and spin-down electrons in the sample.

3.3.1 Electron Exchange in the (e,2e) Scattering on a Ferromagnetic Surface

To investigate experimentally the exchange interaction, one should monitor the scattering of, at least, two electrons with well-defined spin projections and then flip the spin projection of one of the electrons in which case the symmetry of the spin part of the wave function is changed. To obtain a clear picture on the influence of this symmetry operation one should resolve all the relevant quantum numbers of the two electrons before and after the collisions (Berakdar 1999 PRL). Experimentally this

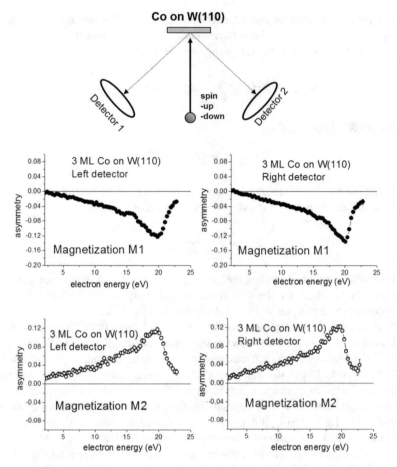

Fig. 3.10 Asymmetry of SPEELS recorded in left and right detectors for magnetization of the sample "up" (first row) and "down" (second row)

is a challenging task since the coincident two-electron detection combined with the measurement of the spin-polarization of electrons results in very low counting rates.

A theoretical description of state-resolved scattering in an electronic system requires consideration of the excited-state of a many-body system, which is notoriously difficult. It is necessary to take into account two kinds of correlations. First, it is electronic correlation in the single-particle ground and excited state of the surface. This situation is similar to other widely used single-particle spectroscopies, such as single photoemission and electron-energy loss spectroscopy. The electronic correlation of this kind is included here on the level of density functional theory within the local-density approximation. The second kind of correlation that has to be included in the theory is the interaction between the two electrons escaping into the vacuum.

The approach here is first to choose an appropriate model for the electron–electron interaction and then to evaluate numerically the probability of the electron pair

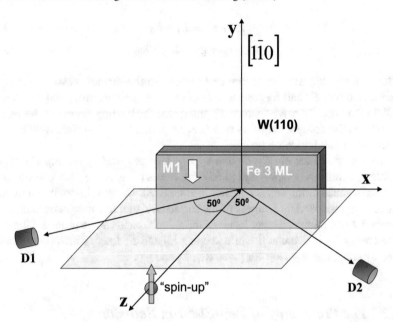

Fig. 3.11 Spin-polarized incident electron beam impinges onto a ferromagnetic surface with magnetic moment **M** collinear with the polarization vector of the beam **P**

emission under electron impact. The theory does not account for multi-step electron-electron scattering of the emitted electron pairs from other electrons of the conduction band. Therefore, the validity of the theoretical model is restricted to the vicinity of the Fermi level E_F. This is because the processes involving multi-step scattering, which are irrelevant at E_F, are strongly spin dependent and are the primly source of the spin decoherence.

For a concise presentation the discussion is restricted initially to the experimental situation studied in (Samarin et al. 2000a PRL). A polarized electron beam with a polarization vector $\mathbf{P}_1$ and a wave vector $\mathbf{k}_0$ impinges onto a clean ferromagnetic surface (here a BCC Fe(110)) with a well-defined magnetization direction $\mathbf{M}$ (cf. Fig. 3.11). The question to be addressed is the following: How does the emission rate of the two electrons change if the direction of $\mathbf{P}_1$ or/and $\mathbf{M}$ is inverted? In other words, the quantity of interest is a spin asymmetry A which is determined by measuring the electron pair emission rate I when the magnetization direction (hereafter denoted by $\Downarrow$) is antiparallel or parallel to the polarization vector ($\downarrow$) of the incident electron beam. The asymmetry A is then evaluated as:

$$A(\mathbf{k}_0, \mathbf{k}_1, \mathbf{k}_2) = \frac{I(\uparrow\Downarrow) - I(\downarrow\Downarrow)}{I(\uparrow\Downarrow) + I(\downarrow\Downarrow)} \tag{3.63}$$

In the above electron-electron scattering the following energy and momentum conservations apply:

$$E_0 + E_b = E_1 + E_2 \tag{3.64}$$

$$\mathbf{k}_{0\parallel} + \mathbf{q}_\parallel + \mathbf{g}_\parallel = \mathbf{k}_{1\parallel} + \mathbf{k}_{2\parallel} \tag{3.65}$$

Here E_0 and $\mathbf{k}_{0\parallel}$ are the energy and parallel to the surface wave vector of the incident electron. E_1 and E_2 are energies of two outgoing electrons; $\mathbf{k}_{1\parallel}$ and $\mathbf{k}_{2\parallel}$ are parallel to the surface wave vectors. E_b and $\mathbf{q}_\parallel$ are the binding energy of the valence electron and the component of its wave vector parallel to the surface, and $\mathbf{g}_\parallel$ is the surface reciprocal lattice vector.

Since the quantities E_0, E_1, E_2 and $\mathbf{k}_0$, $\mathbf{k}_{1\parallel}$, $\mathbf{k}_{2\parallel}$ are given by the experiment it is possible to control, via (3.64) and (3.65), the values of E_b and $\mathbf{g}_\parallel$, i.e. one can zoom into certain states in the surface Brillouin zone (BZ). Using spin-polarized incident electrons, spin-dependent features of the surface BZ of a ferromagnetic sample can be probed. For example, deeper levels of the valence band can be probed if the value of the energy E_2 is decreased while keeping E_0 and E_1 fixed. Similarly, $\mathbf{g}_\parallel$ can be varied by changing, e.g., $\mathbf{k}_{0\parallel}$ for given $\mathbf{k}_{1\parallel}$, $\mathbf{k}_\parallel$ and $\mathbf{g}_\parallel$.

3.3.2 The Probability of Two-Electron Emission from a Ferromagnetic Surface Under Spin-Polarized Electron Impact

The incident spin-polarized electron beam can be described by the density operator ρ^{s_1} with matrix elements $\rho^{S_1}_{m_{S_1} m_{S_1}}$, where m_{S_1} is the projection of the electron's spin s_1 . The density operator can be linearly expanded in terms of Pauli matrices σ (Berakdar 2002 Nucl Inst):

$$\rho^{S_1} = 1 + \mathbf{P} \cdot \sigma. \tag{3.66}$$

For the description of the valence electrons, one can employ the density matrix $\overline{\rho}^{S_2}_{m_{S_2} m_{S_2}}$, where s_2 is the spin of the valence electron and m_{S_2} stands for the spin projection. The representation of the density operator $\overline{\rho}^{s_2}$ is obtained from:

$$\overline{\rho}^{S_2} = w_0(\mathbf{q}_\parallel, l, E_b)(1 + \mathbf{P}_2 \cdot \sigma) \tag{3.67}$$

Here $w_0(\mathbf{q}_\parallel, l, E_b)$ is the spin-averaged Bloch spectral function of the layer l and $\mathbf{P}_2$ is the spin polarization of the valence electron state, i.e.:

$$P_2 = \frac{w(\mathbf{q}_\parallel, l, E_b, \Downarrow) - w(\mathbf{q}_\parallel, l, E_b, \Uparrow)}{w_0(\mathbf{q}_\parallel, l, E_b)} \tag{3.68}$$

The quantities $w(\mathbf{q}_\parallel, l, E_b, \Downarrow)$ and $w(\mathbf{q}_\parallel, l, E_b, \Uparrow)$ are the Bloch spectral functions of the majority and the minority bands, respectively, whereas w_0 is the spin averaged Bloch spectral function.

The spectral density functions of the sample can be obtained from conventional surface band structure calculations (Blaha et al. 1990; Qian and Hübner 1999). The cross section W for the simultaneous emission of two electrons with wave vectors $\mathbf{k}_1$ and $\mathbf{k}_2$ in response to the impact of a projectile electron with wave vector $\mathbf{k}_0$ is given by (Berakdar 2002 Nucl Inst):

$$W(\mathbf{k}_0, \mathbf{k}_1, \mathbf{k}_2) = C \sum_{m'_{S_1} m'_{S_2} m_{S_1} m_{S_2}} \int d\alpha \sum T \rho^S T^\dagger \delta(E_f - E_i), \qquad (3.69)$$

where $C = (2\pi)^4/k_0$, and $\rho^S = \rho^{S_1} \otimes \bar{\rho}^{S_2}$.

Spin-orbit interactions in the scattering process are neglected here. This model does not account for energy-loss processes of the electron pair because they are not allowed at the Fermi level according to the energy conservation requirement (3.64), however if electronic levels deeper in the band are involved one has to consider these multiple-step processes. Hence, the above sketched model is justifiable for the electron-pair emission near the vicinity of E_F but it becomes less applicable if the valence electrons involved in the collisions are far below the E_F.

Here the asymmetry function A is defined as follows (Berakdar 2002 Nucl Inst):

$$A = P_1 \frac{\sum_l [w(\mathbf{\Lambda}_\|, l, \varepsilon, \Downarrow) - w(\mathbf{\Lambda}_\|, l, \varepsilon, \Uparrow)] \sum_{g_\|} X_{av} A^S \delta(E_f - E_i)}{\sum_{l'} w_0(\mathbf{\Lambda}_\|, l', \varepsilon) \sum_{g'_\|} X_{av} \delta(E_f - E_i)} = \frac{W(\uparrow\Downarrow) - W(\downarrow\Downarrow)}{W(\uparrow\Downarrow) + W(\downarrow\Downarrow)}$$

$$(3.70)$$

The wave vector $\mathbf{\Lambda}_\|$ is introduced here:

$\mathbf{\Lambda}_\| = \mathbf{K}_\|^+ - \mathbf{g}_\| - \mathbf{k}_{0\|}$, where $\mathbf{K}_\|^+ = \mathbf{k}_{1\|} + \mathbf{k}_{2\|}$, and A^S is the "exchange scattering asymmetry" defined as:

$$A^S \equiv \frac{X^{(S=1)}(\mathbf{k}_1, \mathbf{k}_2; \mathbf{k}_0, \mathbf{g}_\|, l) - X^{(S=0)}(\mathbf{k}_1, \mathbf{k}_2; \mathbf{k}_0, \mathbf{g}_\|, l)}{3X^{(S=1)}(\mathbf{k}_1, \mathbf{k}_2; \mathbf{k}_0, \mathbf{g}_\|, l) + X^{(S=0)}(\mathbf{k}_1, \mathbf{k}_2; \mathbf{k}_0, \mathbf{g}_\|, l)} \qquad (3.71)$$

3.3.3 Connection Between Measured Asymmetry of the (e,2e) Spectrum and Spin-Dependent Electronic Structure of the Target

In the highly symmetric experimental arrangement in which correlated electrons are detected at equal angles and equal energies the triplet cross-section $X^{(S=1)}$ is zero. It follows that $A^s = -1$ and the measured asymmetry takes on the form (Berakdar 2002 Nucl Inst):

$$A = -P_1 \frac{w(\mathbf{\Lambda}_\|, \varepsilon, \Downarrow) - w(\mathbf{\Lambda}_\|, \varepsilon, \Uparrow)}{w(\mathbf{\Lambda}_\|, \varepsilon, \Downarrow) + w(\mathbf{\Lambda}_\|, \varepsilon, \Uparrow)}, \qquad (3.72)$$

where P_1 is the degree of polarization of the incident beam. In other words, the measured asymmetry A is directly related to the spin asymmetry of the spectral density function $w(\mathbf{\Lambda}_\|, \varepsilon, \Downarrow\Uparrow)$ of the valence band.

These very symmetric experimental conditions are obtained only when electron pairs are detected with equal energies and at equal angles ($E_1 = E_2$, and $\theta_1 = \theta_2$). This condition corresponds to the excitation of the valence electron from the Γ point of the surface Brillouin zone. Indeed, if $\mathbf{k}_{1v} = -\mathbf{k}_{2\parallel}$ and $\mathbf{k}_{0\parallel} = 0$, then $\mathbf{q}_{\parallel} = 0$ as follows from (3.64) and (3.65). The different values of $E_1 = E_2$ correspond to different binding energies of the valence electron at the Γ point of the surface Brillouin zone:

$$E_b = E_1 + E_2 - E_0. \qquad (3.73)$$

In order to have reasonable statistics and satisfy symmetry requirements ($E_1 = E_2$, and $\theta_1 = \theta_2$) one can proceed as follows. First select narrow bands on the detectors such that $\theta_1 \approx \theta_2$ (See Fig. 3.13) and then select a band along the diagonal on the 2D energy distribution of correlated electron pairs, where $E_1 \approx E_2$ (See Fig. 3.12a, b and between two white lines). For each combination of angles (θ_i, θ_i) the asymmetry of the spectral density function (SDF) can be calculated in the Γ point of the surface

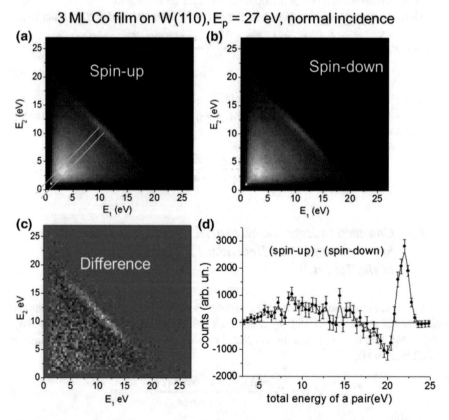

Fig. 3.12 Energy distributions of correlated electron pairs excited by spin-polarized electrons from a ferromagnetic Co layer on W(110). **a, b** 2D energy distributions for spin-up and spin-down incident electrons; **c** difference of (**a**) and (**b**); **d** difference of binding energy spectra for spin-up and spin-down incident electrons

Fig. 3.13 Selection of
symmetric stripes on the
detectors D1 and D2

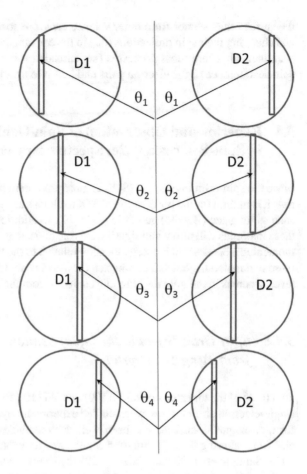

Fig. 3.13 Selection of symmetric stripes on the detectors D1 and D2

BZ as a function of the binding energy. Then these asymmetries for different angles can be averaged (See Fig. 3.13).

To compare the measured asymmetry of the spectral density function with a calculation, the theory needs to take into account the finite detection solid angles.

In an experiment, electrons are detected within solid angle Ω_θ, with an aperture Δ_θ. The sampling over Ω_θ implies an averaging over $\mathbf{k}_1$ and $\mathbf{k}_2$ within a certain range. From (3.65) it follows that this procedure corresponds to integration in a certain region in the BZ around the Γ point. Since at the symmetric conditions the asymmetry A reflects the asymmetry of the spectral density function (SDF) it should not depend on the primary electron energy E_0. This statement is endorsed by experimental and theoretical results for the asymmetry near the Fermi energy. It should be noted that the theory performs well near the Fermi level because the approximations made in the theory are justifiable near E_F as pointed out above. However, if the theory is applied beyond its range of validity, i.e. away from E_F then considerable deviations between experiment and theory are observed. This is due to the fact that, when the emission occurs from deeper levels in the band, the two electrons before escaping the surface may scatter from other valence electrons losing some of their energies and changing their spin projections (due to the exchange interaction). A sequence of such inelastic energy-loss processes (i.e. cascade process) may lead to a complete spin decoherence of the electron pairs and hence to a vanishing spin asymmetry.

3.4 Experimental Observation of Spin-Orbit Effects in Spin-Polarized (e,2e) Spectroscopy on Surfaces

Since the spin-orbit interaction (SOI) is most prominent in the case of large-Z materials it is natural for the observation of SOI to choose as the sample large-Z materials such as tungsten ($Z = 74$) and gold ($ZZ = 79$). In addition, a comparison of SOI in these materials will allow highlighting the role of d-states (with large orbital angular momentum) because their energy location relative to the Fermi energy is different in these two materials. Indeed, in tungsten the d-states are located just below the fermi level, whereas in the gold they are moved down from the Fermi level by about 2 eV.

3.4.1 Spin-Orbit Effect in the (e,2e) Scattering from Tungsten Crystal

Two faces of the tungsten crystal W(100) and W(110), for which there are extensive single-electron observations, were studied experimentally. The spin-orbit interaction for non-magnetic surfaces has been well demonstrated experimentally for single electron scattering from the tungsten crystal. See for example, (Kirschner and Feder 1979; Samarin et al. 2005b, 2007a, b, 2013b; Venus et al. 2000).

The polarization direction of the incident electron beam is chosen perpendicular to the scattering plane containing the incident beam and two detectors. In such a geometrical arrangement, one can expect in general a non-vanishing spin-asymmetry in measured spectra because of symmetry considerations (Gollisch et al. 1999). The [100] direction of the tungsten crystal is perpendicular to the scattering plane in both cases of W(100) and W(110) and the sample can be rotated around this axis varying the angle of incidence θ. Since the accumulation of data with satisfactory statistics takes approximately 30 h the measurements were stopped every few hours and the sample was cleaned by a high temperature flash to remove adsorbed residual gases even though the base pressure in the vacuum chamber was in the low 10^{-11} Torr range.

Energy and momentum conservation of the (e,2e) reaction imply that the valence electron involved in the collision with a projectile can be localized in energy-momentum space. As above pointed out, in a single electron collision of the incident electron with a valence electron the energy of the primary electron E_1 and two outgoing electrons E_3, E_4 define the binding energy of the valence electron: $E_2 = E_3 + E_4 - E_1$. The number of correlated electron pairs as a function of the total (sum) energy

$E_{tot} = E_3 + E_4$ then represents the "total energy distribution" or as a function of the binding energy, a "binding energy spectrum". For a fixed total energy, the two electrons of the pair can share this energy in different ways depending on the scattering dynamics and electronic properties of the surface. The number of electron pairs as a function of the difference energy $(E_3 - E_4)$ presents an "energy sharing distribution" for a given total energy, or within a narrow band of total energy. In the case of a crystal surface, the parallel-to-the-surface component of the electron momentum is a good quantum number for four relevant electronic states of the scattering event. In other words: $\mathbf{k}_{1\parallel} + \mathbf{k}_{2\parallel} = \mathbf{k}_{3\parallel} + \mathbf{k}_{4\parallel}$, where $\mathbf{k}_1, \mathbf{k}_2, \mathbf{k}_3, \mathbf{k}_4$ are the momenta of the incident, bound, first-detected and second-detected electrons, respectively. Hence, one can present the measured spin-polarized (e,2e) spectra as projections on the binding energy, on $\mathbf{k}_{2\parallel}$, or as energy sharing distributions within a certain total energy band.

Figure 3.14 represents binding energy spectra measured at 25.5 eV primary energy for three different orientations of the sample: normal incidence (panel *b*) and two off-normal (panels *a* and *c*). The angle of incidence was changed by rotating the sample around the axis perpendicular to the scattering plane by ± 12°. It is noted that the shape of the binding energy spectrum depends on the incident angle. At normal incidence, there are two maxima in the spectrum: one at about 1 eV below the Fermi level and the second at about 7 eV below the Fermi level. In contrast, for both off-normal (±12°) positions of the sample the binding energy spectrum contains a single pronounced maximum located at 1 eV below the Fermi level. Regarding the spin dependence, the normal incidence spectra (panel b) for "spin-up" and "spin-down" primary beam are identical and the difference spectrum is zero. For off-normal incidence the "spin-up" and "spin-down" spectra are different for both positions (a) and (c) and difference spectra have opposite signs: negative for the sample position (a) and positive for the position (c). The difference between spin-up- and spin-down spectra is located in the energy range within a few eV below the Fermi level. It was shown (Artamonov et al. 1997; Feder et al. 1998) that in this energy range the major contribution to the (e,2e) spectrum comes from the single step two-electron collisions whereas for lower energies the contribution from multi-step collisions is substantial.

The dynamics and kinematics of individual electron-electron scattering are exploring how two correlated electrons share their energy within 2 eV total (binding) energy just below the Fermi level. For normal incidence the energy sharing distributions from W(100) and W(110) are shown in Fig. 3.15. First, note that the shape of the distribution for W(100) is different from that measured for W(110). Secondly, the differences between spin-up and spin-down spectra for both faces of the tungsten crystal are different and exhibit an interesting symmetry feature as predicted by theory (Gollisch et al. 1999). Mirror reflection with respect to the $E_1 - E_2 = 0$ point transforms I^+ almost perfectly into I^- and hence changes the sign of the difference spectrum. This result is consistent with the theoretical results of (Gollisch et al. 1999; Feder et al. 1998) and can be understood from the symmetry analysis of the experimental geometry (Figs. 3.7 and 3.10). Reflection at the (y, z)-plane reverses the spin of the incident electron (because the spin is an axial vector) and interchanges the two outgoing electrons. It implies that $I^+(E_3 - E_4) = I^-(E_4 - E_3)$.

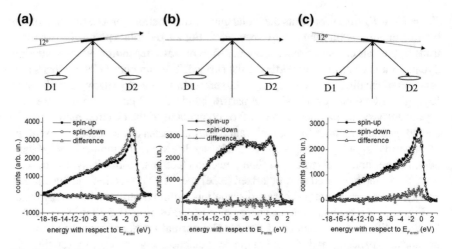

Fig. 3.14 Binding energy spectra of W(110) recorded at three orientations of the sample and primary electron energy 25.5 eV. (Reprinted figure with permission from Samarin et al. (2004), Copyright (2004) by APS)

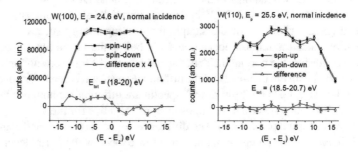

Fig. 3.15 Energy sharing distributions of correlated electron pairs from W(100) and W(110) recorded at normal incidence. (Reprinted figure with permission from Samarin et al. (2004), Copyright (2004) by APS)

When the incidence angle changes from zero to $+12°$ or to $-12°$ the energy sharing distributions change dramatically, both in shape and in spin-dependence (Fig. 3.16). For normal incidence the distribution is symmetric with respect to the zero point where $E_1 = E_2$. For off-normal incidence (panels a and c) the distributions are not symmetric relative to the zero point. For the sample position (a) there is a maximum at $E_1 - E_2 = -10$ eV whereas for the position (c) the maximum is located at $E_1 - E_2 = 10$ eV. If we take the medium total energy $E_{tot} = 19.5$ eV, then the two maxima correspond to the following combinations of electron energies: ($E_1 = 4.75$ eV; $E_2 = 14.75$ eV) for the sample position (a) and ($E_1 = 14.75$ eV; $E_2 = 4.75$ eV) for the sample position (c). The combinations of momenta would be ($K_1 = 1.12$ Å^{-1}, $K_2 = 1.97$ Å^{-1}) and ($K_1 = 1.97$ Å^{-1}, $K_2 = 1.12$ Å^{-1}), respectively. Assuming for the sake of simplicity that electrons are detected in the centers of the detectors, the total momentum of the pair would be close to the direction of the specularly reflected primary beam.

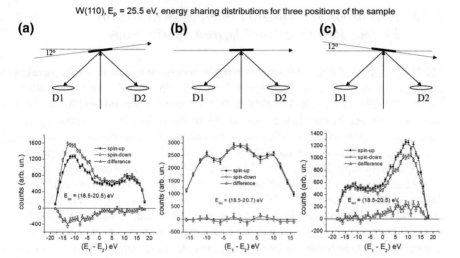

Fig. 3.16 Energy sharing distributions of correlated electron pairs excited by 25.5 eV primary electrons from W(110) at three different positions of the sample and [100] axis perpendicular to the scattering plane. (Reprinted figure with permission from Samarin et al. (2004), Copyright (2004) by APS)

This means simply that the total momentum of the pair carries the parallel component of the incident electron momentum because of the momentum conservation law. It is interesting to note that the average coincidence count rate for off-normal incidence is about two times higher than for normal incidence. Together with the arguments presented in the discussion of the binding energy spectra for normal and off-normal incidence this indicates that the off-normal incidence is more favorable for observing correlated two-electron emission resulting from the binary collision of the incident electron with the valence electron. Regarding the spin-dependence, it is seen that the sharing distributions for off-normal incidences exhibit large differences between spin-up and spin-down spectra. The difference spectra possess broad maxima located at $(E_1-E_2)=-10$ eV for positions (a) and at $(E_1-E_2)=10$ eV for position (c). The asymmetry $A = (I^+-I^-)/(I^++I^-)$ reaches -10% and 10%, respectively. The I^+ and I^- spectra and difference spectrum $D = (I^+-I^-)$ show again an interesting symmetry property. Here I_a and I_c denote spectra recorded at geometry (a) and geometry (c), respectively. Reflection at the (y,z)-plane reverses the spin of the primary electron and transforms the geometry (a) into geometry (c) with interchange of the outgoing electrons. It implies that $I_a^+(E_1-E_2) = I_c^-(E_2-E_1)$ and, by consequence, $D_a(E_1-E_2)=-D_c(E_2-E_1)$. Comparing the spectra in panels (a) and (c) it is seen that difference spectra exhibit such symmetry.

3.4.2   Anisotropy of Spin-Orbit Interaction in W(110) by Spin-Polarized Two-Electron Spectroscopy

The (110) surface of a tungsten single crystal is often used as a substrate for growing model metal films, as its free energy is high and inter-diffusion or surface alloying usually does not occur. A variety of ferromagnetic layers on W(110) surfaces have been studied to establish a relationship between the films' morphology and their magnetic properties including magnetic anisotropy. One particular example is a ferromagnetic layer of iron on W(110) surface that exhibits a thickness dependent magnetic anisotropy: for thin layers (thickness <40–50 ML) the easy magnetic axis is directed along the [1$\bar{1}$0] direction in the plane of the film, whereas thick films (thickness >50 ML) have the [100] easy magnetic axis as in a bulk material (Donath et al. 1999). This in-plane rotation of the easy magnetic axis of the Fe/W(110) film when its thickness reaches a critical value was extensively studied by various techniques (Elmers and Gradmann 1990; Fruchart et al. 1999; Sander 2004). The origin of magnetic anisotropy, in general, is the dipole-dipole interaction (shape anisotropy) and spin-orbit coupling (magnetocrystalline anisotropy) (Bruno 1993). In ultrathin ferromagnetic layers in addition to the shape anisotropy there are two important contributions to the magnetic anisotropies: Néel-type magnetic interface anisotropy originating from symmetry breaking spin-orbit coupling at the interface (Elmers and Gradmann 1990) and magnetoelastic anisotropy resulting from an epitaxial strain (Sander 2004). Since the interface anisotropy includes the interaction between atoms of the film and the substrate its contribution to the total anisotropy may be altered by changing the substrate or by depositing a cap layer of foreign atoms. The role of surface anisotropy in a spin reorientation transition (SRT) in Fe(110) thin films on W(110) was studied using spin-resolved photoelectron spectroscopy (Baek et al. 2003). The thickness dependence of SRT was discussed in terms of a competition between surface and volume contributions to the total energy of the system. In a thin film the surface anisotropy favors the in-plane [1$\bar{1}$0] easy magnetization axis whereas in a thick film (thickness > critical thickness t_r) the bulk anisotropy, which favors in-plane [001] magnetization axis, prevails and the easy magnetization switches from the [1$\bar{1}$0] to the [001] direction. It was demonstrated (Baek et al. 2003) that the balance between surface and bulk anisotropy can be changed by modifying the surface of the iron film. In fact, the critical thickness t_r was substantially reduced by depositing 1 or 2 monolayers of Ag or Au on top of Fe film and the t_r reduction was considerably more pronounced with Au than with Ag over layers. These results demonstrate the importance of at least one of the surface anisotropies (Fe/vacuum) contribution to the magnetic anisotropy of thin ferromagnetic films.

The contribution of the film/substrate interface was shown, contrary to the case of Fe/W(110) films, that the Fe/Mo(110) interface anisotropy favors the in-plane magnetization alignment along the [001] axis even for a very thin Fe layer (about 10 ML). It is interesting to note that the lattice constant of a Mo crystal is almost identical to that of W ($a_{Mo} = 0.3147$ nm, $a_W = 0.31652$ nm). It means the lattice constant mismatch between the substrate and the ferromagnetic film and, by consequence, the strain in the film, seems to be similar in these two cases. However, the magnetic anisotropies of Fe/W and Fe/Mo are different. On the other hand the iron films deposited on GaAs crystal (with the lattice mismatch of only 1.3%, compared to about 10% in cases of W and Mo) exhibit the same behavior as on W(110) (Prinz et al. 1982).

Ab initio magnetocrystalline anisotropy calculations for Fe/W(110) and Fe/Mo(110) systems (Qian and Hübner 2001) showed that in the case of W there is a much stronger interface anisotropy than in the case of Mo despite the similarities of their mechanical, atomic and electronic properties. It was suggested (Qian and Hübner 2001) that the reason is the fact that W is a heavier atom than Mo. The W 5d electrons are further away from the nuclei compared to the Mo 4d electrons and, by consequence, the 5d orbitals are more influenced by and influence more strongly the neighboring atoms of the deposited Fe film. The spin-orbit interaction for the W 5d electrons is larger than for the Mo 4d ones. Since the origin of the magnetic anisotropy is the anisotropy of the crystal lattice (Bruno 1989) together with spin-orbit coupling, the magnitude of the orbital moment and its anisotropy directly reflect the anisotropy of the crystal environment and the distortion due to spin-orbit coupling (Qian and Hübner 2001; Ravindran et al. 2001). Concerning the lattice misfit between the Fe layer and the W and Mo substrates (11.6% and 10.5%, respectively), the layer of Fe might be a little bit more stretched on the W than on the Mo substrate and this may influence the magnetic anisotropy through magnetoelastic coupling. On the other hand, theoretical results (Qian and Hübner 2001) on thin Fe free-standing films show very little dependence of the magneto-crystalline anisotropy energy on the lattice parameters variation if the symmetry and b/a ratio of the film lattice are fixed. The magnetoelastic effect is shown to be a very complicated function of the film electronic structure.

Magnetic anisotropy is correlated with the anisotropy of the orbital moment (Bruno 1989; Ravindran et al. 2001). This correlation suggests that the easy magnetization direction is the direction of largest orbital moment (Bruno 1993; Ravindran et al. 2001). Since the bonding between Fe atoms and W substrate most likely occurs via the direct d-d interaction (Qian and Hübner 2001), one can assume that if there is a strong orbital anisotropy of the d states of W(110) they will impose the orbital anisotropy onto d orbitals of the Fe layer through the hybridization. This will influence the magnetic anisotropy. Therefore it is a challenging task to visualize the orbital anisotropy in the surface of W(110) using spin-orbit interaction as a probing tool.

It appears that the spin-polarized two-electron coincidence spectroscopy in reflection geometry is an efficient instrument for studying anisotropy of spin-orbit coupling in W(110). Experimental results on the spin-polarized two-electron coincidence spectroscopy of a W(110) surface reveal an anisotropy of the spin-orbit interaction on this surface and, by consequence, the anisotropy of the electron orbital moment.

To reveal the difference in the spin-dependent electron scattering dynamics for two inequivalent scattering planes measurements were performed for two azimuthal orientations of the sample shown in Fig. 3.17a, b. Figure 3.17c, d show the energy sharing distributions for two positions of the sample and for two spin polarizations of the incident beam. First, the overall shapes of the sharing distributions are different for the two positions of the sample. For the position (a) there is a maximum in the center of the distribution, corresponding to the equal energies of the two electrons $(E_1 = E_2)$, and two shoulders at $(E_1 - E_2) \approx \pm 15$ eV. For the position (b) there is a minimum at $E_1 = E_2$ and two shoulders at $(E_1 - E_2) \approx \pm 15$ eV. This difference is not surprising because of the two-fold symmetry of the (110) surface of the W crystal and it may be related to the different diffraction conditions for the correlated electron pairs (Samarin et al. 2000b Surf Sci) and to the different density of states along these two directions in the Brillouin zone (Feder et al. 1998).

Consider the spin-difference spectra and asymmetries. For the sample position (a), when the [100] direction is perpendicular to the scattering plane, there is a minimum in the difference spectrum at $(E_1 - E_2) \approx -10$ eV and the maximum at $(E_1 - E_2) \approx 10$ eV. The corresponding asymmetry spectrum has a minimum value of $A = -5\%$ and a maximum of $A = 5\%$ at about the same difference energies (± 10 eV). In the central part of the spectrum, from $(E_1 - E_2) \approx -5$ to $(E_1 - E_2) \approx 5$ eV, the asymmetry is about zero within the arrow bars. For the sample position b), when the [1$\bar{1}$0] direction is perpendicular to the scattering plane, the difference spectrum (Fig. 3.17d) and corresponding asymmetry (Fig. 3.17f) are totally different from the previous case. The difference spectrum has the maximum at $(E_1 - E_2) \approx -9$ eV and the minimum at $(E_1 - E_2) \approx 9$ eV. The asymmetry has the maximum value of about 7.5% and minimum of about –7.5% at the same difference energies of ± 9 eV. The asymmetry changes from maximum value to the minimum almost linearly, and crosses zero at $(E_1 - E_2) \approx 0$. For both positions of the sample the difference spectrum and the asymmetry exhibit typical for spin-orbit coupling "left-right" symmetry that is consistent with the symmetry property of the experimental setup. These results indicate that the spin-orbit interaction depends very much on the azimuthal position of the sample. The reason for that is the two-fold symmetry of the crystal potential on the (110) surface.

An overview of the azimuthal dependence of the spin-effect in spin-polarized (e,2e) spectroscopy of W(110) is shown in Fig. 3.18. Azimuth = 0 ($\varphi = 0$) corresponds to the azimuthal position of the sample, when the [1$\bar{1}$0] crystallographic direction in the surface plane of the sample is perpendicular to the scattering plane. The left panel of the figure shows asymmetries of the SPEELS of W(110) recorded with 27 eV primary energy at normal incidence and detection by two MCP detectors located at $\pm 50°$ with respect to the normal. It means two SPEELS are recorded: (i) by detector D1 and (ii) by detector D2 (see Fig. 3.17b). These distributions demonstrate "left-right" symmetry in spectra that is typical for the spin-orbit interaction. The right panel shows asymmetries of the sharing distributions of the (e,2e) spectra measured with 27 eV primary energy at normal incidence with the azimuthal positions indicated on figures. The total energy of the pairs is chosen to be within 2 eV band just below the Fermi energy. The asymmetry of the sharing distribution (right column) for $\varphi = 0$ has maximal value compared to two other azimuthal positions of the sample. At $\varphi = 45°$ it becomes smaller, and at $\varphi = 90°$ it even changes the sign (similar to Fig. 3.17). In single-electron spectra (left column) asymmetry of inelastic scattering for the sample positions $\varphi = 0$ and $\varphi = 45°$ is larger than for $\varphi = 90°$. However, for elastic scattering ($E = 27$ eV) the asymmetry for $\varphi = 90°$ is larger than

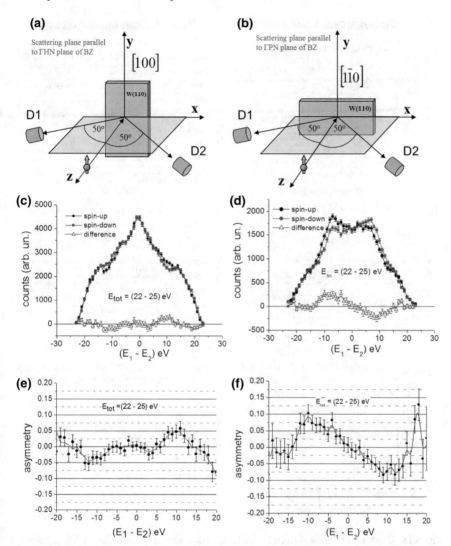

Fig. 3.17 Geometrical arrangement (**a** and **b**), energy sharing distributions (**c** and **d**) and asymmetries of the energy sharing distributions (**e** and **f**) for two azimuthal orientations of the sample. Primary electron energy is 29 eV, normal incidence. (Reprinted figure with permission from Samarin et al. (2005b), Copyright (2005) by APS)

for two other positions of the sample. One can plot a 2-D projection of the (e,2e) distribution on the $(K_x - K_y)$ plane (where K_x and K_y are components of the valence electron wave vector).

The difference of two-dimensional $(K_x - K_y)$ distributions measured with spin-up and spin-down incident electrons is presented in Fig. 3.19. The horizontal line (axis) on the figure is parallel to the scattering plane which, in turn, is parallel to the

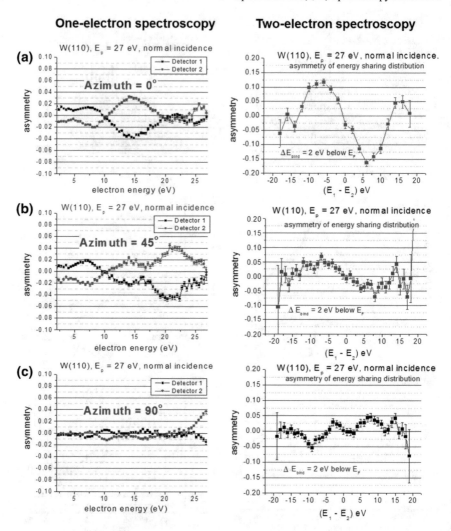

Fig. 3.18 Overview of the spin effects in single- and two-electron spectroscopy of W(110) at various azimuthal positions of the sample. (From Samarin et al. (2011b). Reproduced with permission. All rights reserved)

ΓPN plane of the tungsten Brillouin zone. The particular symmetry of the patterns of the figure shows roughly the two-fold symmetry with respect to the point shifted upwards from the origin of the (k_x, k_y) coordinate system. This shift indicates most likely an imperfection of the sample orientation. The figure illustrates an advantage of the two-dimensional mapping of the distributions compared to the one-dimensional projections. The projection of the distribution (Fig. 3.19) on the k_x axis gives a large difference between spin-up and spiny-down spectra, whereas the projection on the k_y axis will show smaller difference (and asymmetry). This is because for the k_x

Fig. 3.19 W(110), $E_p =$ 27.5 eV, normal incidence. Scattering plane parallel to ΓPN plane of BZ. $K_x - K_y$ map for binding energy 2 eV below $E_{Fermi.}$, difference [(spin-up)–(spin-down)]

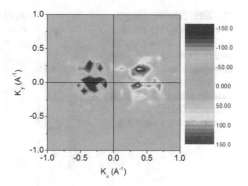

projection the values of the difference have the same sign in positive and negative sides of the k_x axis. In contrast, for the k_y projection the difference values have positive and negative signs for both positive and negative sides of the k_y axis.

3.4.3 Spin-Orbit Effects in the (e,2e) Scattering from Au(111) Film on W(110) Surface

A few atomic layers of gold on a W(110) substrate form a good epitaxial layer with the (111) face parallel to the sample surface. The gold film was deposited onto the W(110) substrate at room temperature by evaporation from a small piece of gold wire melted on a V-shaped tungsten filament heated resistively. The quality of the film was demonstrated by observing quantum well states in such a layer of gold using photoelectron spectroscopy (Shikin et al. 2002) and layer-by-layer growth was confirmed by MEED oscillations (Knoppe and Bauer 1993a, b).

A set of the binding energy curves for Au(111) is shown in Fig. 3.20. A sharp maximum just below the Fermi level, which is very prominent for low primary energy, corresponds to the surface state. The arguments in favor of this suggestion are the following.

First, in the bulk density of states of gold there is a flat and low density of states in the energy range just below the Fermi level down to about 1.5 eV below the Fermi level. Hence, the maximum just below the Fermi level cannot be formed by the electron excitation from the bulk states. Second, its energy position (about 0.5 eV below the Fermi level) corresponds to the calculated and measured (by photoelectron spectroscopy) position of Shockley surface states. The second maximum in the binding energy spectrum at about (2–3) eV below the Fermi level corresponds to the emission of pairs from the valence d-derived states.

Figure 3.20 shows explicitly that the shape of the binding energy spectrum is changing with the primary electron energy. In particular, the relative intensity of the peak just below the Fermi energy decreases while the primary energy increases. It shows that the relative probability of excitation of surface and bulk states depends

Fig. 3.20 Binding energy spectra of Au(111) film on W(110) for various primary energy. Zero point on x-scale corresponds to a position of the Fermi level. (Reprinted from Samarin et al. (2015b), Copyright (2015), with permission from Elsevier)

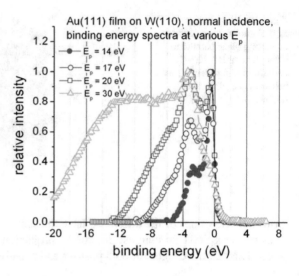

Fig. 3.21 Energy dependence of the relative contribution of surface and bulk states to the binding energy spectrum of Au(111) film on W(110). (Reprinted from Samarin et al. (2015b), Copyright (2015), with permission from Elsevier)

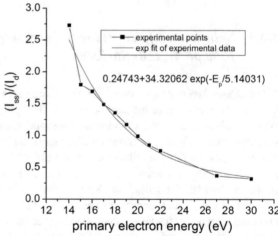

on the primary electron energy. The (e,2e) spectra of Au(111) were measured with increments of 1 eV in the incident electron energy. The relative contribution of surface states into the (e,2e) binding energy spectrum was analyzed by plotting the ratio of the surface—to d-derived bulk valence states intensity as a function of primary energy as shown in Fig. 3.21.

An exponential fit to this data gives the following relationship:

$$(I_{SS}/I_d) = 0.25 + 34.32 \cdot \exp\left(-E_p/5.14\right) \tag{3.74).}$$

It is seen that the relative contribution of surface states in the binding energy spectrum decreases with the increase of the primary electron energy and the ratio span an order of magnitude change in the primary energy range from 14 to 30 eV. The increase of the contribution of correlated pairs from surface states relative to the bulk contribution at low primary energies might be due to a few reasons. First, suppression of the contributions of d-states to the measured (e,2e) spectrum is expected when the primary electron energy drops below 15 eV because those d-states are located about 3.5 eV below the Fermi level and an electron from these states must gain energy of at least 8.5 eV (binding energy plus the work function) to be excited above the vacuum level. If two electrons are excited by 15 eV primary energy and escape the surface, they will share an excess energy of $15 - 8.5 = 6.5$ eV. If they share energy equally they can be detected, in principle, if both of them are within the acceptance angles of the detectors. If either of them is refracted out of the acceptance angle of a detector, the pair can not be detected. If the pair of electrons shares energy unequally and one of them has energy below 2.5 eV it will not be detected because the time-of-flight analyzer was set to detect electrons only with energy greater than 2.5 eV. In summary, d-states with binding energy about -3.5 eV give small contributions to the (e,2e) spectrum at low primary electron energies.

The second reason arises from the diffraction of correlated electron pairs. Indeed, in our geometry and energy range we observe the (00) diffraction pattern of electron pairs (Schumann e al. 2013).

Electrons of pairs excited from the surface state have almost equal energies and their total momentum points along the normal to the surface. The intensity of this (00) pattern is energy dependent. One can assume that, in the studied energy range, the contributions (intensities) of the (00) patterns corresponding to the excitation of pairs from surface states and d-states change in different ways with the change of primary energy. It will result in a substantial change of the relative intensity of the peaks in the binding energy spectrum.

The increase of the contribution of electron pairs in the binding energy spectrum at energies below –8 eV with increase of primary energy is most likely due to the contribution of multi-step electron scattering that generates "true secondary" electrons. Consequently, it means that the incident electron undergoes an inelastic scattering with a valence electron of the target and before these two electrons escape from the surface they can participate in other one or few electron-electron collisions generating low-energy "true secondary" electrons.

To get new insight into the role of the elastic scattering (of the incident electron) in the (e,2e) reaction, the azimuthal position of the Au(111) film was changed in such a way that the diffracted (01) beams were moved into the scattering plane (see Fig. 3.22 top panel). The asymmetries of SPEELS recorded by two detectors, Det. 1 and Det. 2, which are located symmetrically with respect to the normal to the sample surface, changed dramatically (see left and right graphs of Fig. 3.22b). The most significant change of the asymmetry is observed for elastic scattering (elastic maximum). Indeed, at the azimuthal angle $\varphi = 0°$ the asymmetry of elastic scattering is zero, whereas for $\varphi = 30°$ it reaches the values $+5.5\%$ and -5.5% for the left and right detectors, respectively. However, the energy sharing distribution of the (e,2e) spectrum and the asymmetry of the energy sharing distribution (Fig. 3.22c, d) do not change much when the azimuthal angle changes. This is in contrast to the case of a W(110) sample, where the azimuthal rotation of the sample by 90° changed the asymmetry of the sharing distribution, not only the shape but also the sign. The fact that the change of the asymmetry of elastic scattering of primary electrons does not have much influence on the asymmetry of the sharing distribution of the (e,2e) spectrum indicates that the scenario, where the spin-orbit asymmetry in the incident electron scattering determines the asymmetry of the (e,2e) scattering, does not work. The favorable scenario becomes apparent for showing where the spin-orbit interaction in the valence state determines the overall spin-orbit asymmetry of the (e,2e) spectrum.

Both tungsten (Z = 74) and gold (Z = 79) are the large-Z (heavy) elements where the spin-orbit interaction is known to be strong. The question is how the SOI shows up in the (e,2e) reaction and where it is localized in energy-momentum space of the electronic structure of the sample. One of the advantages of the spin-polarized (e,2e) spectroscopy is the possibility to localize bound electron taking part in the (e,2e) reaction in energy-momentum space of the valence band. To find the binding energy at which the SOI is most prominent, slices of constant total energy of pairs are compared with analysis of the asymmetry of the energy sharing distributions. Figure 3.23 shows a two-dimensional distribution, with two spin-resolved energy sharing distributions for two bands of total energy measured on Au(111) film on W(110) with normal incidence of 19 eV spin-polarized primary electrons.

One can see on the two-dimensional distribution of correlated electron pairs (Fig. 3.23b) two ridges along fixed total energies of pairs: the first ridge is just below the offset of the distribution which corresponds to the position of the Fermi level at $E_{\text{tot}} \approx 14.2$ eV. This offset means that the top most energy level, from which the valence electron can be excited in the (e,2e) reaction, is the Fermi level. Note that not taken into account is the thermal tail of the Fermi-Dirac distribution of the valence electrons. The second ridge is located at $E_{\text{tot}} \approx 11$ eV which corresponds to the energy position of the d-derived valence states of the gold. These two ridges are better visualized in the total energy distribution presented in Fig. 3.23a). The narrow one with the maximum located at $E_{\text{tot}} = 13.75$ eV corresponds to the pairs excited from the surface states of Au(111), whereas the second one at $E_{\text{tot}} = 11$ eV corresponds to the pairs excitation from the d-derived states. It is seen that the shapes of the sharing distributions (Fig. 3.23c and d) are different.

Figure 3.24 shows the asymmetries of the sharing distributions for the same total energies of the correlated electron pairs. It is obvious from the figures that the spin asymmetry of the sharing distribution, taken at the total energy of pairs corresponding to the pairs excitation from the d—states of the gold valence band, is larger than that from the surface states of the Au(111) sample. This finding highlights the role of the spin-orbit interaction in the valence state involved in the (e,2e) scattering and indicates that the valence states with larger orbital momentum (d-states compared to s-p surface states) result in a larger spin-orbit asymmetry of the (e,2e) spectrum. Note that the overall spin-orbit asymmetry of the (e,2e) spectrum is determined by the spin-orbit interaction in all relevant states involved in the (e,2e) reaction, i.e. the incident electron state (E_1, $\mathbf{k}_1$), valence state (E_2, $\mathbf{k}_2$), and two outgoing electron states (E_3, $\mathbf{k}_3$) and (E_4, $\mathbf{k}_4$). Then it is interesting to check to what extent the spin-

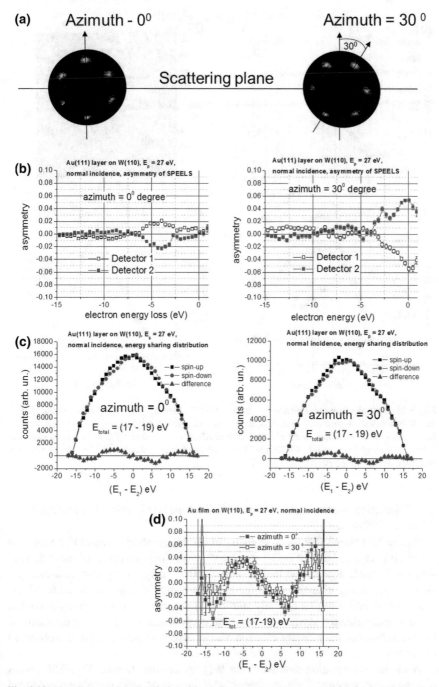

Fig. 3.22 How the sample azimuthal position influences the asymmetry of the sharing distribution of correlated electron pairs excited from Au (111) film on W(110) by spin-polarized 27 eV electrons at normal incidence

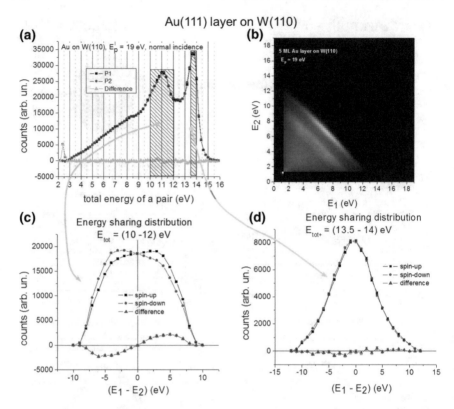

Fig. 3.23 Two-electron coincidence spectra of Au(111) film on W(110) surface measured with 19 eV primary spin-polarized electrons at normal incidence; **a** total energy distribution; **b** spin-integrated two-dimensional distribution; **c** spin-resolved energy sharing distribution of electrons with total energy in the band (10–12) eV; **d** spin-resolved energy sharing distribution of electrons with total energy in the band (13.5–14) eV

orbit interaction in other states is responsible for the whole spin-orbit asymmetry of the (e,2e) spectrum.

Figure 3.25 shows that the asymmetry of the energy sharing distribution taken at the binding energy band of $(-3.2 \div -4$ eV), where the maximum of the d-derived states is located, depends on the primary electron energy. It appears that even the sign of asymmetry changes when the incident energy changes from 22 to 27 eV. This finding indicates that not only spin-orbit interaction in the valence state of the sample determines the asymmetry in the (e,2e) spectrum (in the sharing distribution, for example), but the primary electron energy also is responsible for the observed asymmetry.

A similar observation was made on W(110) sample. Indeed, Fig. 3.26 shows binding energy spectra (Fig. 3.26a) recorded at the normal incidence of spin-polarized electrons on W(110) surface with incident energy 13 and 27 eV. It is obvious from this figure that incident electrons with the energy of 13 eV can excite the valence electrons

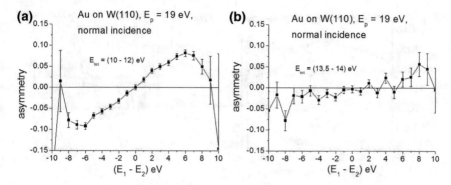

Fig. 3.24 Asymmetries of the energy sharing distributions for two different total energy bands of the (e,2e) spectrum of Au film on W(110) excited by 19 eV spin-polarized electrons at normal incidence

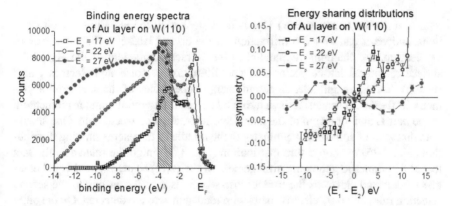

Fig. 3.25 Binding energy spectra (left) and the asymmetries of energy sharing distributions (right) for Au(111) film on W(110) excited by primary electrons with different energies. The shaded rectangle defines the binding energy band

only from energy levels within the energy band of few eV below the Fermi level. A black rectangular on the figure indicate the energy band of the spectrum, where the K_x-distributions were analyzed. These distributions are shown in Fig. 3.26b. Similar to the case of Au(111) (Fig. 3.25) the shape and even the sign of the distributions corresponding to different incident electron energies are different.

All these allow suggesting the following scenario of spin-orbit effect in the (e,2e) scattering. An incident electron with its orbital angular momentum and particular spin state interacts with a valence electron with its orbital angular momentum and spin state. The electron pair resulting from such an interaction is characterized by the total momentum (total wave vector) $\mathbf{K}_{tot} = \mathbf{k}_3 + \mathbf{k}_4$, total orbital angular momentum $\mathbf{L} = \mathbf{L}_1 + \mathbf{L}_2$ and total spin of the pair $\mathbf{S}$ with projection on quantization axis equal to $S_{1,2} = 0$ or $S_{1,2} = 1$, i.e. singlet or triplet scattering channel. It was shown (Samarin et al. 2000b SS) that the concept of a quasiparticle of the pair with the

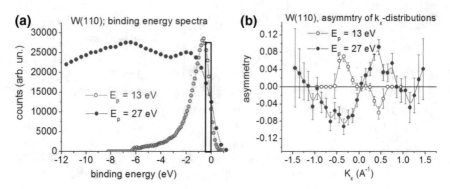

Fig. 3.26 Binding energy spectra (**a**) and K_x-distributions (**b**) of W(110) measured for two incident electron energies: $E_p = 13$ eV and $E_p = 27$ eV. (From Samarin et al. (2011b). Reproduced with permission. All rights reserved)

total momentum (wave vector) $\mathbf{K}_{tot}$ is consistent with the experimental observations. Indeed it can undergo diffraction on the crystal lattice in the same way as an electron, i.e. changing its parallel to the surface component of the $\mathbf{K}_{tot\parallel}$ by a surface reciprocal lattice vector $g_{\parallel}$ (see "diffraction of correlated electron pairs" section). So, the magnitude, sign, and energy dependence of the spin-orbit asymmetry in the (e,2e) scattering is determined by the total orbital angular momentum of the pair $\mathbf{L}$ and total spin of the pair $\mathbf{S}$ and interaction between them. One should note that zero of spin-orbit asymmetry in the middle of the energy sharing distribution (Fig. 3.25) or momentum distribution (Fig. 3.26) might be related to the fact that at a symmetric geometric arrangement of the (e,2e) scattering and equal energies of detected electrons the triplet cross-section is zero (according to the second selection rule) and only electron pairs with total spin zero are detected. On the other hand this "quasi-particle" with zero spin projection does not experience a spin-orbit interaction, i.e. spin-orbit asymmetry is zero.

This scenario, we suggest, is supported by results on W(100) surface with off-normal incidence (see Figs. 3.14 and 3.16). They show that the spin-orbit asymmetry depends not only on the face of the crystal, azimuthal position of the sample, and energy of the incident electrons, but also on the polar angle of incidence. For different polar angles of incidence the ratio between singlet and triplet channels of scattering should be different for each combination of energies within electron pair as well as total angular momentum of the pair. These change spin-orbit interaction within a pair and result in a different shape of the spin-orbit asymmetry in a sharing distribution.

3.4.4   Comparison of the Spin-Polarized (e,2e) Scattering from W(110) and Au(111)

Comparison of the (e,2e) spectra and asymmetries measured on W(110) surface and on Au(111) thin film on W(110) shows that the maximum asymmetry of the sharing

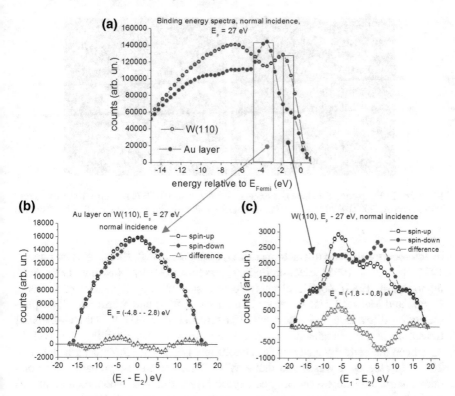

Fig. 3.27 Comparison of (e,2e) spectra of W(110) and Au(111) recorded at $E_p = 27$ eV, normal incidence; **a** binding energy spectra; **b** spin-resolved energy sharing distribution of Au(111); **c** spin-resolved energy sharing distribution of W(110). (From Samarin et al. (2011b). Reproduced with permission.

distributions is observed at the total energy of pairs corresponding to the excitation of a valence electron from the d-derived states of the valence band (see Fig. 3.27).

This observation confirms again that the valence states with large spin-orbit interaction, that is the valence states derived from the atomic states with large orbital angular momentum, make a large contribution to the observed SOI asymmetry of the (e,2e) spectrum.

3.4.5 Influence of the Oxygen Adsorption on Spin-Polarized Two-Electron Scattering from W(110)

The tungsten surface and oxygen adsorption on W(110) are the most studied model systems of surface science. They have been studied using most of the surface science techniques. The geometrical and electronic structures have been characterized

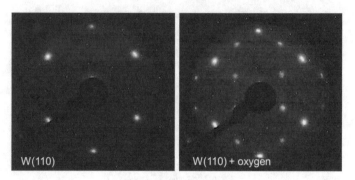

Fig. 3.28 LEED picture for clean and oxygen covered surface of W(110), Ep = 100.8 eV. (From Williams et al. (2007) 012024. © IOP Publishing. Reproduced with permission. All rights reserved)

by low-energy electron diffraction (LEED) (Tringides 1990; Buchholtz and Lagally 1974; Baek et al. 1993), Electron Energy Loss Spectroscopy (Rawlings 1980; Artamonov et al. 1985), core-level photoelectron spectroscopy (Ynzunza et al. 2000; Fuggle and Menzel 1975; Peden and Shinn 1994; Riffe and Wertheim 1998), photoelectron diffraction and holography (Ynzunza et al. 2000a, b). Order—disorder transitions as well as adatom—adatom interactions in an oxygen layer on W(110) have been theoretically analysed (Buchholtz and Lagally 1975; Zaluska-Kotur et al. 2002). In essence it was found that a W(110) surface can be prepared clean and unreconstructed whereas the ordered oxygen layer exhibits several structural phases depending on the exposure and temperature treatment (Wu et al. 1989).

Much less is known regarding details of the modification of the electronic structure of W(110) upon oxygen adsorption. Core-level shift induced by oxygen adsorption and indicating of a strong oxygen-tungsten interaction was observed using photoelectron spectroscopy (Fuggle and Menzel 1975; Peden and Shinn 1994; Riffe and Wertheim 1998). Even less information is available on the modification of the valence states of W(110) upon oxygen adsorption (Peden and Shinn 1994; Riffe and Wertheim 1998; Opila and Gomer 1981). The main modification due to the oxygen adsorption in the photoelectron spectrum from the valence band of the W(110) is the appearance of a broad maximum approximately 6 eV below the Fermi level. It was attributed to the emission from O(2p)-derived levels of the system. Some variation of intensity in the energy range from O(2p) band up to the Fermi level was also reported (Peden and Shinn 1994; Riffe and Wertheim 1998).

Consider an oxygen ordered layer formed at 1300 °C sample temperature and exposure to oxygen at a pressure 2×10^{-8} Torr for 4 min. Figure 3.28 presents LEED patterns for clean and oxygen covered W(110) surfaces. It can be seen that oxygen atoms form $p(2 \times 2)$ structure (with probable presence of the $p(2 \times 1)$ domains Wu et al. 1989).

The energy conservation law in the (e,2e) reaction allows the extraction of the binding energy of the valence electron, E_b from measured energies of two correlated electrons: $E_b = (E_1 + E_2) - E_o$, where E_o is the incident electron energy, E_1 and E_2 are the two detected electrons energy. It has been shown previously (Artamonov et al. 1997) that single-step scattering is the primary process in the total energy region 3–4 eV below the Fermi edge ($E_b = 0$). Below this energy interval a multi-step scattering dominates and the momentum-energy-conservation equations are no longer applicable. Within a total energy band ΔE_{tot} where the single-step processes are dominant the energy sharing distribution would show how two correlated electrons share the total energy of the pair. In the case of a single crystal target, the electron momentum parallel to the surface is conserved and one can trace the bound electron momentum: $\mathbf{k_{b\parallel}} = \left(\mathbf{k_{1\parallel}} + \mathbf{k_{2\parallel}}\right) - \mathbf{k_{o\parallel}}$, where $\mathbf{k_o}$ is the incident electron momentum, $\mathbf{k_1}$ and $\mathbf{k_2}$ are the two detected electron momenta. We note here that measurements of the (e,2e) distributions collected under these conditions do not directly represent the density of states as a function of the binding energy or the momentum density distribution. This is because in the low-energy (e,2e) reaction the cross-section depends on the Coulomb matrix elements and on the reflectivities of the primary and ejected electrons at the surface (Gollisch et al. 1999), as well as on the energy- and angular-dependence of the dynamics of scattering. On the other hand, to obtain a significant scattering cross section an appreciable density of states in momentum space is necessary (Gollisch et al. 1999). In other words, if upon oxygen adsorption a maximum appears in the binding energy spectrum it would indicate, most likely, an increase of the density of states at a corresponding binding energy.

Figure 3.29a, c present binding energy spectra I(E_b) of clean and oxygen covered W(110) surfaces recorded at normal incidence with 27 eV primary energy electrons. In both cases spin-up and spin-down spectra are identical. The main difference between spectra of clean tungsten and oxygen covered W(110) surfaces is a broad maximum in the spectrum of the oxygen covered tungsten (Fig. 3.29c) at the binding energy of 6 eV. It indicates most likely an increase of the density of states at this energy because each value of the binding energy spectrum results from integration of the cross section over a range of both electrons energies along the total energy band. Therefore, dynamical features are largely smeared out. This maximum is due to the 2p-orbital of oxygen that establishes the bonding between oxygen atom and the W(110) surface and consequently increases the density of state at the surface in this energy range. A similar feature in the binding energy spectrum of an oxidized tungsten surface was observed in the earlier study of oxygen adsorption on W(001) by (e,2e) spectroscopy (Samarin et al. 1999). Photoelectron spectroscopy also reveals this maximum in the W(110)/O system (Feydt et al. 1999) and also the $O(2p)$–derived orbital centred at 6 eV binding energy showed drastic change in the $W(5d)$ region induced by the oxygen adsorption (Feydt et al. 1999). The new features in the photoemission spectrum in this energy range (close to the Fermi level) clearly show a two-dimensional character. No reconstruction in the topmost layer of tungsten was detected. This extra structure in the spectrum was interpreted as an emission from $W(5d)$–$O(2p)$ hybridized orbitals energetically split off the bulk bands due to the action of a modified surface potential. In the binding energy spectrum in Fig. 3.29c these new features, due to the oxygen adsorption in the energy range 3 eV below the Fermi level, are not visible. Only the slope of the onset in the clean W spectrum is larger than in the oxidized W spectrum.

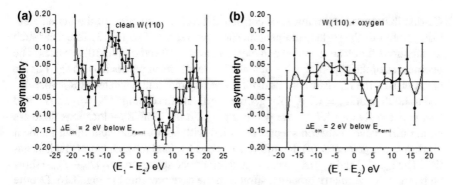

Fig. 3.30 Asymmetries in energy sharing distributions for clean and oxygen covered W(110) surface. (From Williams et al. (2007). © IOP Publishing. Reproduced with permission. All rights reserved)

adsorption. Another reason for that might be the change of the surface barrier at the oxidized surface. A qualitative discussion of this effect follows.

As stated above the major contribution to the measured spin-orbit asymmetry in a tungsten (e,2e) spectrum is related to the SOI in valence states (Gollisch et al. 1999). In addition, the shape of the surface barrier also plays an important role in the escape process of the two outgoing electrons. They undergo refraction and experience spin-orbit interaction while traversing the barrier. At this stage a spin-filter effect may occur as was observed for the photoemission from nonmagnetic material (Oepen et al. 1986) and for the secondary emission excited by spin-polarized electrons from W(110) (Samarin et al. 2005a, b SS). Indeed, the scattered and ejected electrons might have opposite spin projections on the quantization axis (polarization of the incident beam) due to a dominant singlet scattering in this energy range and therefore experience different (spin-dependent) potential barriers while travelling through the surface. It results in a spin-filter effect, i.e. electron transmission through the surface depends on the spin state of the electron. The oxygen adsorption modifying the surface barrier influences (reduces) the spin-filter effect.

A strong influence of the oxygen adsorption on the emission of correlated electron pairs from W(110) can be seen in K_x-distributions (Fig. 3.31). In this experiment the sample was rotated around the normal to the surface such that $[1\bar{1}0]$ direction was in the scattering plane containing the sample normal and two detectors: let's call it "position 2" in contrast to "position 1" when the $[1\bar{1}0]$ direction is perpendicular to the scattering plane. It was shown that at normal incidence and for this azimuthal orientation of the sample the spin-orbit asymmetry in the sharing distributions (see Fig. 3.17) is weaker than for "position 1" (Samarin et al. 2005b PRB), but when the sample was tilted (polar angle was changed) towards detector 1 or detector 2 the strong asymmetry shows up in the binding energy spectra as well as in the sharing distributions (Samarin et al. 2004 PRB). Then, the very significant advantage of the position sensitive detection acknowledges the direct identification of the measured

K_x-distributions as the number of correlated pairs as a function of the bound electron parallel-to-the surface momentum (or wave vector). Figure 3.31 represent such distributions for clean (a and c) and oxygen covered (b and d) W(110) surfaces. The geometries of the experiment are denoted in the inserts of panels (a) and (c). For the clean surface the K_x-distributions exhibit a strong maximum located at $K_x = 0.5$ Å^{-1} and a shoulder at $K_x = 1.7$ Å^{-1} for geometry (a) and maximum at $K_x = -0.5$ Å^{-1} and a shoulder at $K_x = -1.7$ Å^{-1} for geometry (c). At the same K_x—locations there are strong differences between spectra recorded with spin-up and spin-down incident electrons. For geometry (a) the difference is positive and for geometry (c) it is negative. The $I_{up}(K_x)$ and $I_{dow}(K_x)$ spectra and difference spectrum $D = (I_{up} - I_{dow})$ show an interesting symmetry property, similar to the case presented in Fig. 3.16. Denote by I^a and I^c spectra recorded at geometry (a) and geometry (c), respectively. Reflection at the (y, z)-plane reverses the spin of the primary electron and transforms the geometry (a) into geometry (c) with interchange of the outgoing electrons. It implies that $I_{up}^a(K_x) = I_{down}^c(-K_x)$ and, by consequence, $D^a(K_x) = -D^c(-K_x)$. Comparing the spectra in panels (a) and (c) it can be seen that the difference spectra exhibit such symmetry. Adsorption of the oxygen on the surface of the sample changes dramatically the shape of K_x-distributions for both geometries. Now a strong maximum is located at $K_x = 1.5$ Å^{-1} and a shoulder at $K_x = 0.5$ Å^{-1} for geometry (a) and a maximum at $K_x = -1.5$ Å^{-1} and a shoulder at $K_x = -0.5$ Å^{-1} for geometry (c).

This modification of the shapes of K_x-distributions upon oxygen adsorption cannot be directly related to the change of the density of states distribution as was pointed out above. On the other hand K_x-distributions measured on oxygen covered W(100) surface at very different geometry (Samarin et al. 1999 J Phys France) also show maxima at $K_x = 0.5$ Å^{-1} and $K_x = 1.5$ Å^{-1} for binding energy around 6 eV. The corresponding calculated distribution reproduced reasonably well the experimental curve (Samarin et al. 1999 J Phys France). It is interesting to note that momentum density distributions of the 2p state of free atomic oxygen show a broad maximum centred at $K_b = 1.5$ Å^{-1} (Samarin et al. 1999). Although the K_x-distributions are constructed for binding energy above $O(2p)$ band they show oxygen-like behaviour in the case of oxygen covered W(110) surface. This might indicate the hybridization of $O(2p)$ and $W(5d)$ states and reconstruction of the momentum density distribution. Concerning the spin-dependent asymmetry of the K_x-distributions, the oxygen adsorption almost completely washes out the spin-dependence (Fig. 3.31b, d). The reason for that is the modification of the valence states of the tungsten as well as the surface potential barrier.

3.5   Experimental Observation of the Exchange Effect in (e,2e) Scattering on Ferromagnetic Surfaces

In contrast to the spin-orbit interaction the exchange effect in the scattering of spin-polarized electrons does not depend on the orientation of the spin (polarization vector) of the incident beam relative to the scattering plane but rather depends on the mutual

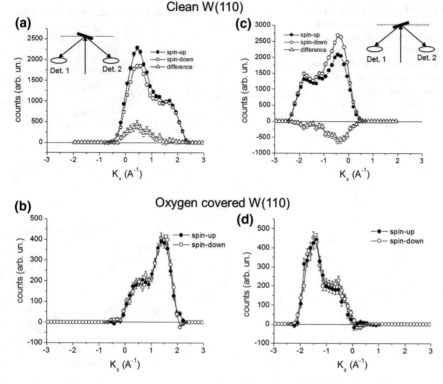

Fig. 3.31 Influence of the oxygen adsorption on the K_x-distributions of clean and oxygen covered surface of W(110). $E_p = 25$ eV, sample was tilted to detector 1 (**a, b**) and to detector 2 (**c, d**) by 12°. The distributions are taken within 2 eV slice of binding energy just below the Fermi level. (From Williams et al. (2007). © IOP Publishing. Reproduced with permission.

orientation of spins of the electrons involved in the collision. To observe the role of exchange in the scattering of electrons it is necessary to control the spin state of at least one of the electrons and perform scattering at two opposite mutual orientations of the spins. In the case of a solid-state sample ferromagnets offer targets with spin - aligned electrons.

The following section presents a few examples of the spin-polarized electron scattering from ferromagnetic samples that includes a single crystal of iron, thin Fe film on W(110), and thin Co film on W(110). In addition, a couple of bi-layered structures, Co/Ni and Au/Fe on W(110) will be analysed. The first one has demonstrated the influence of a buffer (ferromagnetic) layer on the properties of the Co film. The second one shows how the spin-related properties of a surface can be tailored.

Fig. 3.32 SPEELS (**a**) and asymmetry (**b**) measured on Fe(110) at normal incidence of 20 eV primary electrons and 40° detection. (Reprinted from Samarin et al. (2001), Copyright (2001), with permission from Elsevier)

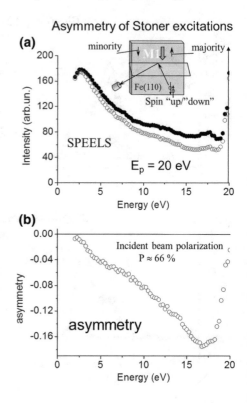

3.5.1 Fe(110) Single Crystal

An iron single crystal Fe(110) was used as a sample for the first spin-polarized (e,2e) scattering experiment (Samarin et al. 2000a PRL). The sample was mounted on a yoke made of a μ-metal with a magnetizing coil for the sample magnetization and that allowed control of the magnetization (magnetic moment) of the sample. The incident beam was transversally polarized with the degree of polarization of about 65%. The magnetic state of the sample and the polarization of the incident electron beam were confirmed by the SPEELS measurements that are shown in Fig. 3.32. Normal incidence and 40° detection angle were used for SPEELS measurements. The broad arrow on the sample in the Fig. 3.32a indicates the direction of the sample magnetic moment ("magnetization M1"). Black arrows show orientation of the majority (opposite to the magnetization direction) and minority (parallel to the sample magnetization) spins of the valence electrons of the sample. As usually in our experiments the incident beam polarization was altered every 5 s during the measurements.

Figure 3.32a shows electron energy loss spectra measure with spin-up (open circles) and spin-down (filled circles) incident electrons. Figure 3.32b represents the asymmetry of SPEELS that is due to Stoner excitations (Kirschner 1985b PRL). The negative sign of asymmetry corresponds to the magnetization M1 shown in the figure

The calculated difference of the densities of states for minority and majority electrons (Fig. 3.33) is compared with the experimentally measured difference of binding energy spectra in Fig. 3.34b. The shape of the measured difference confirms that the minority states have higher than majority density just below the Fermi level down to approximately -2 eV. Further down in binding energy the majority states start to dominate slightly exceeding the minority density of states in the range from -2 to -10 eV. The shape of the difference binding energy spectrum is fairly well reproduced by the difference of DOSs (with opposite sign) for majority and minority electrons (calculated and measured spectra are scaled for the comparison). It is important to note that for the bulk density of states the spin-up (majority) states have higher density than spin-down (minority) states just below the Fermi level within the binding energy range of about 1 eV. However, for the surface density of states the minority states have higher density than the majority in the binding energy range of 1 eV just below the Fermi level.

The sign of the measured difference between spin-up and spin-down binding energy spectra and its comparison with the calculated surface spin-resolved DOS (Fig. 3.33) indicates that the measurements provide information on the top most atomic layer of the sample. Both measured and calculated differences change the sign at about 1.5 eV binding energy. So, the difference between numbers of spin-up and spin-down states is negative in the range of binding energy from 0 to about -1.5 eV and it is positive in the binding energy range from about -1.5 to -4 eV. It is interesting to analyze how electrons excited from the states within these ranges of binding energy share the energy depending on the spin orientation of the incident electron. Figure 3.35 shows differences of two spin-resolved sharing distributions taken at two different binding energies. Both difference spectra (b) and (c) have maximum absolute value in the middle of the distributions, i.e. at the $E_1 = E_2$, but the sign of the difference is positive for one distribution and negative for the second distribution. The role of the exchange scattering in the present experiment is illustrated in Fig. 3.36a–d (Samarin et al. 2000a PRL). In the energy diagram illustrating the electron scattering and shown in Fig. 3.36a the incident electron after scattering escapes from the surface as the fast electron, i.e., $E_1 > E_2$. This process is called the direct scattering and proceeds with an amplitude f. In contrast, as illustrated in Fig. 3.36c, the incoming electron may *exchange* as much energy and momentum with the initially bound electron that it emerges as the slower one ($E_1 < E_2$). This scattering process is precisely the one active in the case of the aforementioned Stoner excitation. It is usually referred to as the exchange process and is quantified with an amplitude g. Intuitively one can expect that $|f| \gg |g|$ for $E_0 \approx E_1$ and $E_1 \gg E_2 \ll E_0$. Our experiment does not resolve the electron spin projections in the final state, i.e., we cannot distinguish between the processes shown in Fig. 3.36a, c. Thus, the coincident rate for antiparallel (Fig. 3.36a) or parallel (Fig. 3.36b) alignment of the spins of the incoming and the bound electron is proportional to $|f|^2 + |g|^2$ and $|f - g|^2$, respectively. That result follows because the processes shown in Fig. 3.36a, c can be distinguished experimentally while the processes shown in Fig. 3.36b, d are experimentally identical, and hence f and g are added coherently.

Fig. 3.35 **a** Difference of spin-resolved binding energy spectra; **b**, **c** difference of spin-resolved energy sharing distributions for two binding energies marked by rectangles on (**a**)

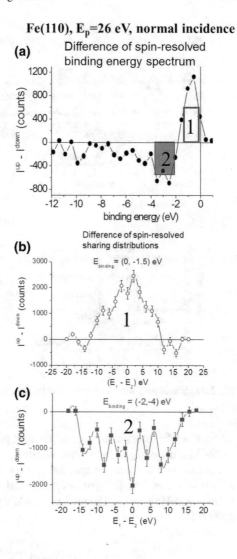

Theoretically, to sum over the (final-state) spin quantum numbers the spins of the electrons are coupled to the (conserved) total spin of the electron pair S to produce two spin channels: the singlet channel ($S = 0$) and the triplet channel ($S = 1$). The singlet and triplet cross section $X^{(S=0)}$ and $X^{(S=1)}$ can then be expressed in terms of f and g.

For a gaseous atomic target with a defined spin polarization P_a (taken as a quantization axis) the asymmetry A is directly expressible in terms of $X^{(S=0/1)}$ (and hence in terms of f and g) as $A = P_e \cdot P_a \cdot X^{(s)}$, where $X^{(s)} = (X^{(S=1)} - X^{(0)})/(3X^{(S=1)} + X^{(S=0)})$ $= (2|f| \cdot |g| \cdot \cos(\delta))/(|f|^2 + |g|^2 + |f + g|^2)$, where δ is the phase difference between the amplitudes f and g. This relation for A implies that $\lim_{|g|/|f| \to 0} A = 0$ reflecting the dynamics of spin-dependent scattering on the polarized atomic target (Streun et al.

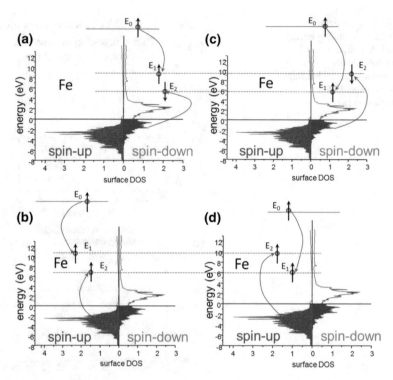

Fig. 3.36 Visual presentation of the direct (**a** and **b**) and exchange (**c** and **d**) scattering channels

1999). For a ferromagnetic surface, the spin polarization of electronic states depends on the binding energy ε and wave vector $\mathbf{q}_\parallel$. As pointed out in Sect. 3.3 the theory has to take into account the spectral density function $w(\mathbf{q}_\parallel, l, E_b, \Downarrow)$ rather than the density of states. The influence of the crystal structure on the scattering dynamics is encompassed in f and g. The *exchange* induced asymmetry A has the form (Berakdar 1999 PRL):

$$A(\mathbf{k}_1, \mathbf{k}_2; \mathbf{k}_0) = P_e \frac{\sum_{l, \mathbf{g}_\parallel} A_l^{(m)} A_{l, \mathbf{g}_\parallel}^{(s)} B_{l, \mathbf{g}_\parallel}}{\sum_{l', \mathbf{g}_\parallel'} B_{l', \mathbf{g}_\parallel'}}. \tag{3.75}$$

Here, the atomic layers parallel to the surface are indexed by l. In (3.75) $A^{(m)}$ describes the *sample's magnetic asymmetry*, whereas the dynamical aspects of the spin dependent collisions are contained in the *exchange scattering asymmetry* $A^{(s)}$. The spin-averaged intensity is referred to as B. This interpretation follows from the definitions of $A^{(m)}$, $A^{(s)}$, and B:

$$A_l^{(m)} = \frac{w(\mathbf{q}_\parallel, l, \varepsilon, \Downarrow) - w(\mathbf{q}_\parallel, l, \varepsilon, \Uparrow w)}{w_0(\mathbf{q}_\parallel, l, \varepsilon)}, \tag{3.76}$$

$$A_{l,\mathbf{g}_\parallel}^{(s)} = \frac{X^{(S=1)}(\mathbf{k}_1, \mathbf{k}_2; \mathbf{k}_0, \mathbf{g}_\parallel, l) - X^{(S=0)}(\mathbf{k}_1, \mathbf{k}_2; \mathbf{k}_0, \mathbf{g}_\parallel, l)}{3X^{(S=1)}(\mathbf{k}_1, \mathbf{k}_2; \mathbf{k}_0, \mathbf{g}_\parallel, l) + X^{(S=0)}(\mathbf{k}_1, \mathbf{k}_2; \mathbf{k}_0, \mathbf{g}_\parallel, l)}, \tag{3.77}$$

$$B_{l,\mathbf{g}_\parallel} = w_0(\mathbf{q}_\parallel, l, \varepsilon)\left[\frac{3}{4}X^{(S=1)} + \frac{1}{4}X^{(S=0)}\right], \tag{3.78}$$

where $w_0(\mathbf{q}_\parallel, l, \varepsilon, \Downarrow)$ and $w_0(\mathbf{q}_\parallel, l, \varepsilon, \Uparrow)$ are the Bloch spectral functions of, respectively, the majority and the minority bands, respectively. Note here that the arrows $\Downarrow$ and $\Uparrow$ indicate the directions of the magnetic moment of the ferromagnetic sample, which is opposite to the direction of the majority spins electron. The spin averaged Bloch spectral function is w_0.

The calculation scheme for $A^{(m)},\ A^{(s)}$, and B is demonstrated in (Berakdar 1999 PRL). For example, $A^{(m)}$, characterising the material, can be derived from band structure calculations within the 'scalar relativistic full potential linearized augmented plane wave' method, for example (Qian and Hübner 1999). $A^{(s)}$ and B are calculated from the layer dependent transition matrix elements.

Equations (3.76–3.78) demonstrate the versatile potential of the pair emission technique for material and scattering dynamics studies: (a) In case of unpolarized electrons and provided $X^{(S=1)}$ and $X^{(S=0)}$ are sufficiently known, the Bloch spectral functions w_0 can be mapped using (3.78). This is documented in (McCarthy and Weigold 1991; Kheifets et al. 1998) for diverse systems; (b) the magnetic asymmetry $A^{(m)}$ [i.e., (3.76)] in the spin-split band structure can be visualized by using polarized electron beam and choosing a geometrical arrangement under which the triplet channel is closed (Berakdar 1999 PRL) ($X^{(S=1)} = 0$) in which case $A^{(s)} = -1$ [cf. (3.77)]; (c) conversely in the case when the spin polarized band structure is known, e.g., from reliable ab initio calculations, $A^{(m)}$ can be deduced from (3.76) and the spin scattering dynamics, which is embedded in $A^{(s)}$, can be extracted from the measured asymmetry A (3.75).

The present experimental setup does not yet allow the exploration in full detail of all these facets of the pair emission technique. In particular, the averaging over the present angular resolution involves integration over $\mathbf{q}_\parallel$ that extends basically over the entire surface Brillouin zone. The $\mathbf{q}_\parallel$ integration of the Bloch spectral functions [cf. (3.76)] yields the surface spin split density of states $\rho(\varepsilon, \Downarrow)$ and $\rho(\varepsilon, \Uparrow)$ that are depicted in Fig. 3.33. For the interpretation of the data we employ thus $A^{(m)} = [\rho(\varepsilon, \Downarrow) - \rho(\varepsilon, \Uparrow)]/[\rho(\varepsilon, \Downarrow) + \rho(\varepsilon, \Uparrow)]$.

In Fig. 3.37a, b the intensity and the asymmetry A are scanned as a function of the energy sharing within the electron pair for a fixed total energy $E_{\text{tot}} = E_1 + E_2$. According to the energy conservation the energy of the valence band state ε is then fixed (E_0 is constant and $\varepsilon = E_1 + E_2 - E_0$). When the two electrons escape with equal energies $E_1 = E_2$ the triplet scattering $X^{(S=1)}$ vanishes due to symmetry (Berakdar 1999 PRL), and hence for $E_1 = E_2$ we obtain $A^{(s)} = -1$, as was demonstrated also experimentally (Streun et al. 1999) (in atomic collisions). Therefore, in this situation ($E_1 = E_2$), the *magnitude* and *sign* of the asymmetry A are dictated merely by $A^{(m)}(\varepsilon)$ which in Fig. 3.37b amounts to $A^{(m)} \approx 20\%$ at $\varepsilon \approx -5$ eV. This interpretation can be

Fig. 3.37 a Spin-resolved
energy sharing distribution;
b Asymmetry of the energy
sharing distribution
(measured and calculated).
(Reprinted from Samarin
et al. (2001), Copyright
(2001), with permission
from Elsevier)

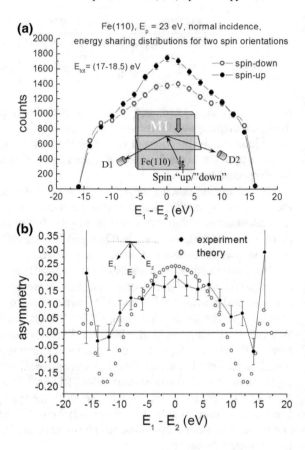

substantiated experimentally by shifting ε deeper into the band [cf. Fig. 3.35b, c] in which case $A^{(m)}$ changes sign.

Since ε is constant (determined by a fixed total energy of pairs) in Fig. 3.37 (and hence $A^{(m)}$ is constant) the variation of A is due to the spin-dependent scattering dynamics which is dictated by $A^{(s)}$. Thus, the structure of A, as depicted in Fig. 3.37b, can be understood as follows: At $E_1 = E_2$ the triplet cross section vanishes in which case $|A^{(s)}|$ attains its highest value (unity). This structure is a peak (minimum) when $A^{(m)} > 0$ ($A^{(m)} < 0$). The decline in A for $E_1 > E_2$ or $E_1 < E_2$ is due to a dominance of the direct scattering amplitude $|f|$ (Fig. 3.36a, b) over the exchange one $|g|$ (Fig. 3.36c, d); i.e., it is more likely for the fast incoming electron to escape as the fast electron than for it to lose almost its whole energy and emerge as the slower one. As deduced above $\lim_{(|g|/|f| \to 0)} A^{(s)} = (|f| \cdot |g| \cdot \cos\delta)/(|f|^2 \cdot |g|^2 + |f| \cdot |g| \cdot \cos\delta) \to 0$, and hence the asymmetry in Fig. 3.37b decreases with increasing deviations from $E_1 = E_2$ (Samarin et al. 2000a PRL).

Figure 3.37b shows the measured asymmetry of the energy sharing distribution at the fixed total energy of pairs $E_{tot} = (E_1 + E_2) = (17 \div 18.5)$ eV. The theoretical curve was calculated using a surface electronic band structure (surface spin-resolved

Fig. 3.38 a The measured (full dots) and calculated (solid curve) asymmetry for $E_0 = 23$ eV, $E_{tot} = 17.7 \pm 0.7$ eV; the sample is tilted with respect to the incident beam by an angle $\alpha = 5°$, as shown in the inset. The calculations (solid curve) are averaged over the angular and energy resolution of the experiment. **b** The same situation as in (**a**) but asymmetry is calculated for different angles α (from $\alpha = 1°$ to $\alpha = 5°$, as depicted on the curves). The excess energy is $E = 18$ eV. The curves are averaged only over the angular resolution of the detectors. (Reprinted figure with permission from Samarin et al. (2000a), Copyright (2000) by APS)

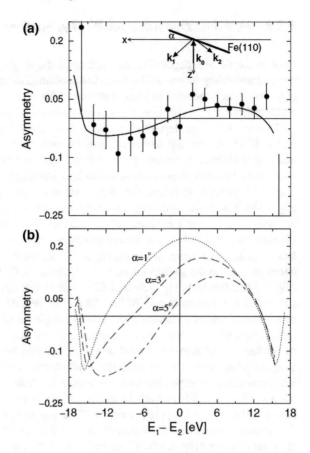

DOS, Fig. 3.33) and it fits the experimental curve fairly well. If the model employs bulk spectral functions it gives the negative sign of asymmetry. This finding confirms that the (e,2e) spectrum provides information on the top most atomic layer of the surface.

The special symmetry (equal detection angles and normal incidence) of the experimental arrangement depicted in Fig. 3.37 implies a symmetrical A with respect to $E_1 = E_2$ (in our case, spin orbit effects are negligibly small). This symmetry is broken by tilting the sample as shown in the inset of Fig. 3.38. It results in the breaking of symmetry of the $A(E_1 - E_2)$ curve (see Fig. 3.38a). Since ε is fixed, $A^{(m)}$ has a fixed constant value in Fig. 3.38. Therefore, the structure (shape) of A is related to that of $A^{(s)}$. It is confirmed by the corresponding calculations (see Fig. 3.38b).

Now let's consider exchange effects in the (e,2e) scattering of spin-polarized electrons from thin ferromagnetic layers of iron and cobalt on a single crystal of W(110).

3.5.2 Fe Films Deposited on W(110) Surface

Prior to the Fe film deposition, the surface of the tungsten crystal was thoroughly cleaned according to the well-established routine (Zakeri et al. 2010; Cortenraada et al. 2001), that includes oxygen treatment of the surface followed by high temperature flashes.

The iron films were deposited on a W(110) surface using an OMICRON type evaporator (EFM-3). The thickness of the film was estimated using a quartz microbalance and Auger electron spectroscopy. The cleanliness and crystal structure of the substrate and deposited films is controlled by integrated LEED-Auger system (SPECS). The LEED pattern taken just after deposition at 119 eV electron energy is shown in Fig. 3.39a. The diffuse spots represent the mean position of the pattern of the Fe(110) structure. A very gentle annealing of the film (about 450 K) leads to the sharp pattern of multiplets arranged symmetrically around the reflections of the bulk Fe(110) surface (Fig. 3.39b). Further annealing (up to 600 K) results in the sharp LEED pattern corresponding to the superposition of the patterns of W(110) and Fe(110) surfaces (Fig. 3.39c). Similar to Fig. 3.39 b LEED structure was observed in such Fe thickness range also in (Gradmann and Waller 1982). The W(110) crystal substrate mounted on a two-fold rotatable manipulator allowed sample rotation around both the vertical (polar angle θ) and horizontal (azimuthal angle ϕ) axes. The $[1\bar{1}0]$ direction along the surface of the tungsten crystal was identified and the azimuthal position of the sample corresponds to $\phi = 0$ when this crystallographic axis runs along the vertical rotational axis and perpendicular to the scattering plane. The magnetic anisotropy of the iron film on W(110) drives the magnetization of the film (in the thickness range below 100 Å) along the $[1\bar{1}0]$ direction of the substrate parallel to the film surface. The substrate was oriented azimuthally such that the $[1\bar{1}0]$ direction was perpendicular to the scattering plane, which spanned through the two detectors and the normal to the sample surface. To define the magnetization (direction of the magnetic moment) of the film, "up" or "down" the deposition of the film was carried out in magnetic field. A weak magnetic field was applied "up" or "down" around the vacuum chamber using a coil during the film deposition. After the deposition magnetic field was turned off and the measurements were performed in a "field-free" space. The strength of the field of about 600 mG was enough to define the direction of magnetization: the direction of magnetization of the film followed the direction of applied magnetic field. When the measurements for a given magnetization were completed, the ferromagnetic film was removed from the substrate by a high temperature flashes. After that a new ferromagnetic layer was deposited in the presence of an external magnetic field, which determined the required direction of the film magnetization. This weak (600 mG or even lower) magnetic field was enough for determining the magnetic moment of the Fe film at the initial stage of the film deposition because the Curie temperature (depending on the film thickness) at that stage was presumably below the room temperature. Being magnetized at the initial stage of growth the Fe film keeps its magnetization direction during the growth of the film while the Curie temperature increases (with thickness) above the room temperature. Hence the film

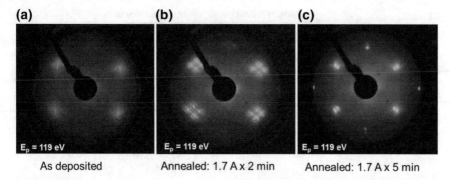

Fig. 3.39 LEED patterns for 5 ML Fe film: **a** as deposited; **b** annealed: 1.7 A × 2 min (about 450 K); **c** annealed: 1.7 A × 5 min (about 600 K). (Reprinted from Samarin et al. (2013a) Copyright (2013), with permission from Elsevier)

remains in remnant magnetization when the film growth is completed and external magnetic field switched off. It is necessary to note that this is not always the case. It may happen that the film undergoes spin reorientation transition while growing, i.e. it changes the direction of magnetization at a certain thickness of the film.

An alternative way to change the orientation of the magnetic moment of the film to its opposite direction is to rotate the sample around the normal to the surface (azimuthal angle) by 180°. It is possible in the case when the vacuum is good and the measurements do not take too long, hence one can assume that the surface of the film is still clean and the contamination would not hamper the measurements at the opposite orientation of the magnetic moment.

Before starting the coincidence measurements, usually the magnetic state of the surface and polarization of the electron beam were checked to ensure a single domain magnetic layer and a proper polarization of the incident beam. In order to do so, the Stoner excitation asymmetry was measured using SPEELS. Figure 3.40 shows asymmetries of SPEELS spectra measured on the Fe(110) for two opposite magnetizations of the sample. Such measurements allow identification of the real orientation of the incident electrons polarization vector if the magnetization direction of the sample is known. Indeed, the asymmetry is defined as (Kirschner 1985a, b):

$$A = \frac{I^{\mathrm{par}} - I^{\mathrm{antipar}}}{I^{\mathrm{par}} + I^{\mathrm{antipar}}},$$

where I^{par} is parallel orientation of the polarization vector of the incident beam and the majority spin orientation of the sample, and I^{antipar} is antiparallel orientation of these vectors. With this definition of asymmetry, the Stoner excitation asymmetry is negative and the corresponding magnetization in our nomenclature is denoted as M1. Of course, if the magnetization of the sample is flipped over to M2 (opposite to M1) and the sequence of "spin-up" and "spin-down" polarization of the incident beam

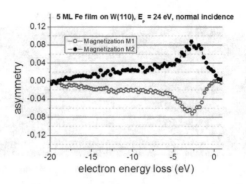

Fig. 3.40 Asymmetry of Stoner excitations for two opposite magnetizations of the Fe(110) film on W(110). (Reprinted from Samarin et al. (2013a), Copyright (2013), with permission from Elsevier)

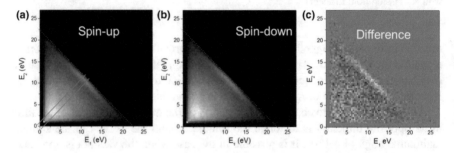

Fig. 3.41 Two-dimensional energy distribution of correlated electron pairs for spin-up (**a**), spin-down (**b**) incident electrons, and the difference spectrum (**c**). The sample is 5 ML Fe film on W(110)

is kept the same as in the previous example, then the sign of asymmetry changes as shown, for example for the Fe film in Fig. 3.40.

The nonzero measured Stoner excitation asymmetry indicates that the film is magnetized and the incident beam is polarized. However, the magnitude of asymmetry indicates the degree of polarization of the electron beam and shows to what extent the film is a single domain structure and how clean is the surface. After such check the coincidence spectrum is measured.

The overview of the (e,2e) spectrum, the two-dimensional energy distribution of correlated electron pairs, is shown in Fig. 3.41. There is a ridge along the cutoff of the spectrum at the total energy of about 22.5 eV (shown in Fig. 3.41a, b by dotted lines), which corresponds to the excitation of the pairs just below the Fermi energy. The difference spectrum has the main contrast right along this ridge.

Binding energy spectra recorded with spin-up I^{up} and spin-down I^{down} incident electrons for two opposite magnetizations M1 and M2 of the sample are shown in Fig. 3.42. The binding energy is shown with respect to the Fermi level. The difference spectra (open circles) have opposite signs for M1 and M2 indicating magnetic origin of the spin dependence.

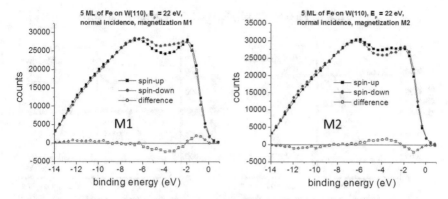

Fig. 3.42 Binding energy spectra recorded with spin-up and spin-down incident electrons for two opposite magnetizations M1 (left) and M2 (right) of the Fe film

The maximum difference between spin-up and spin-down spectra occurs just below the Fermi level ($0 - 2$ eV of binding energy) as well as in the binding energy range between 2 and 6 eV below E_F. This difference is determined by the spin-dependent valence density of states (i.e. energy distribution of the majority and minority electron states) as well as by spin-dependence in the cross section of the correlated electron pair emission under spin-polarized electron impact.

To check how likely the valence state inside the first Brillouin zone (BZ) with various $\mathbf{q}_\parallel$ are excited via (e,2e) reaction, one can fix the total energy of pairs ($E_1 + E_2$) by taking the slice of the 2D energy distribution of pairs, or by taking events within a band of the binding energies. For example, one can take a slice from 0 to -1.5 eV and using relation $\mathbf{k}_{0\parallel} + \mathbf{q}_\parallel + \mathbf{g}_\parallel = \mathbf{k}_{1\parallel} + \mathbf{k}_{2\parallel}$ with $\mathbf{g}_\parallel = 0$ (first BZ) to calculate the number of pairs excited from the valence state with the parallel to the surface crystal momentum $\mathbf{q}_\parallel$ within a given binding energy band.

In the simplest case (one dimensional), one can carry analysis for q_x instead of $\mathbf{q}_\parallel$. Figure 3.43 shows the distributions of number of pairs as a function of q_x for excitation by polarized electrons with spin-up and spin-down from a ferromagnetic film with magnetization $M1$ and $M2$. The distributions are very symmetric with respect to the zero point $q_x = 0$ and the maximum of the distribution is located in the centre of the BZ at $q_x = 0$. The difference spectrum ($I^{up} - I^{dow}$) and asymmetries A_{M1} and A_{M2}, which are defined as:

$$A_{M1} = \left(I^{upM1} - I^{downM1}\right)/\left(I^{upM1} + I^{downM1}\right),$$
$$A_{M2} = \left(I^{upM2} - I^{downM2}\right)/\left(I^{upM2} + I^{downM2}\right)$$

are also symmetric with respect to the zero point $q_x = 0$ (Fig. 3.44). In addition, asymmetries A_{M1} and A_{M2} are almost mirror symmetric with respect to the $A = 0$ line. It means that the spin-orbit contribution to the measured asymmetry is negligible. Indeed, to the leading terms (neglecting interference between exchange and spin-orbit

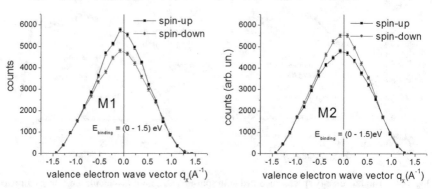

Fig. 3.43 Distribution of correlated electron pairs as a function of the valence electron momentum (wave vector) for spin-up and spin-down incident electrons and magnetization M1 and M2. (Reprinted from Samarin et al. (2015a) with the permission of AIP Publishing)

scattering amplitudes), the spin-orbit and exchange contributions to the measured asymmetry are (Kirschner 1985a, b; Alvarado et al. 1982):

$$A_{ex} = 1/2(A_{M1} - A_{M2})$$
$$A_{SO} = 1/2(A_{M1} + A_{M2}).$$

Figure 3.44b shows that the exchange contribution to the asymmetry of q_x-distribution is almost flat (at the level of about 8%) in the range of the wave vector from -0.5 to 0.5 A^{-1}, and the spin-orbit contribution is zero (Fig. 3.44c) within the error bars in the whole measured range of the q_x.

The asymmetry of spectral density function of the Fe film was measured using symmetric detection of correlated electron pairs as mentioned above and is shown in Fig. 3.45.

3.5.3 Thin Co Films Deposited on W(110)

The substrate W(110) was oriented azimuthally to have the $[1\bar{1}0]$ axis perpendicular to the scattering plane since magnetic anisotropy of a cobalt film drives the magnetization of the film along this direction of the substrate. Prior to the film deposition the W(110) substrate, as always, was cleaned using standard procedure. A cobalt film was deposited on the substrate using a commercial (Omicron-type) evaporator with the deposition rate of a half of monolayer (ML) per minute (Fig. 3.46).

The crystallinity of the 3 ML Co film was checked by low-energy electron diffraction (LEED) that showed a clear sixfold symmetry corresponding to the (0001) face of the *hcp* single crystal of cobalt (Fig. 3.45). The film was magnetized by applying

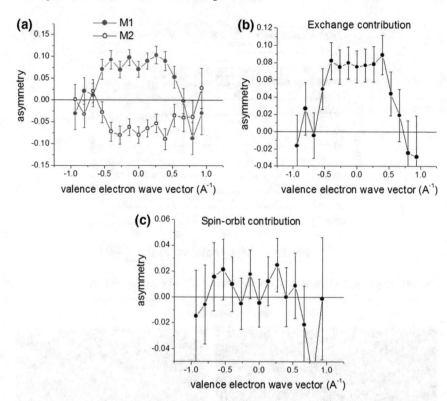

Fig. 3.44 **a** Asymmetry of q_x—distribution measured for magnetizations M1 and M2 of the Fe film; **b** exchange component of the asymmetry; **c** spin-orbit component of the measured asymmetry. (Reprinted from Samarin (2015a) with the permission of AIP Publishing)

400 mG magnetic field during film deposition. The measurements were performed in a field free space after the Earth's magnetic field was compensated down to 3 mG. The base pressure in the chamber was in the 10^{-11} Torr range and rose up to 1.2×10^{-10} Torr during film deposition.

Energy distributions of correlated electron pairs excited by spin-polarized primary electrons of various energies from Co film were measured in the same way as for bulk Fe(110) and for thin Fe films on W(110). Here an analysis is given, in particular of two one-dimensional distributions (the same as for Fe film): for binding energy spectrum and k_x-distribution. The first one relies on the energy conservation law that implies: $E_b + E_0 = E_1 + E_2$, where E_b, E_0, E_1, and E_2 are energies of a bound (valence) electron, primary electron and first and second outgoing electrons, respectively. The second type of the distribution can be constructed in the case of a single crystal sample, when the parallel to the surface component (x-component) of the electron momentum is conserved: $k_{bx} + k_{0x} = k_{1x} + k_{2x} + g_{nx}$, where k_{ix} with i = b, 0, 1, 2 are the x—components (see Figs. 2.10 and 3.10) of the valence electron, incident electron and two outgoing electrons, respectively, and g_{nx} is a reciprocal lattice

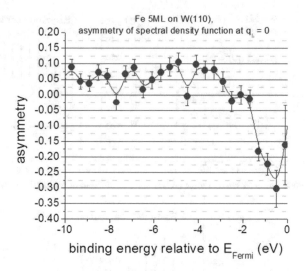

Fig. 3.45 Asymmetry of the spectral density function of the Fe film on W(110)

$$E_p = 143 \text{ eV}$$

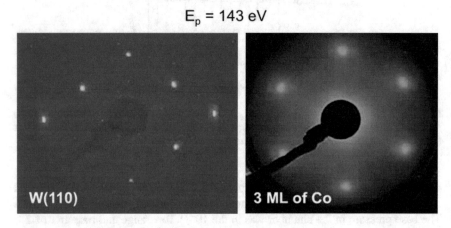

Fig. 3.46 LEED images of clean W(110) and 3 ML of Co on W(110)

vector. Then the binding energy spectrum $I(E_b)$ and the k_x-distribution $I(k_x, E_b = \text{const})$ represent the number of correlated electron pairs as a function of the binding energy or the x-component of the valence electron momentum provided that the second variable of the distributions is constant or confined in a certain range Δk_x or ΔE_b, respectively. Such distributions can be constructed (similarly to the case of Fe film) for two polarizations "up" and "down" of the incident beam: I^{up}, I^{down}, and two magnetization of the film M1 and M2, where superscript "up" and M1 would correspond to the situation shown in Fig. 3.37 (insertion). The polarization "down" means negative projection of the polarization vector on the Y-axis (opposite to the case shown in Fig. 3.37) and M2 corresponds to the magnetic moment of the film opposite to M1. To outline the spin-dependence in these distributions we analyse again asymmetries that are defined as follows:

$$A^{M1,M2}(k_{x1}, k_{x2}, E_0) = \left(I^{upM1,M2} - I^{downM1,M2}\right) / \left(I^{upM1,M2} + I^{downM1,M2}\right). \quad (3.79)$$

Now a discussion is given of the origin of spin-dependence in the measured spectra and how the information on spin-related electronic structure can be extracted from the measured spectra.

Using an experimental array of the (e,2e) spectrum, events can be selected with symmetric detection of electron pairs. Figure 3.47a represents spin-integrated binding energy spectra measured with various primary energies (as indicated on the Fig. 3.47a) and symmetric detection of electron pairs. It can be seen that the spectrum recorded at 9 eV primary energy spans only from the Fermi level down to −1.5 eV binding energy. This is because the low primary energy comparable with the value of the sample work function of about 4.5 eV allows probing only the valence states in the vicinity of the Fermi level. For higher primary energies a larger range of valence states contributes to the measured spectrum.

However, at higher primary energies the low energy contribution of pairs generated in multi-step scattering events becomes substantial. On the other hand, the comparison of spectra for 22 and 27 eV primary energies shows that they are identical. This indicates that at these primary energies the contribution of single-step scattering events is likely dominant in the spectrum within 5 eV binding energies and the spectra reflect, to large extent, the spin-integrated density of states at $k_{bx} = 0$ (Rücker et al. 2005). As an example, the spin-resolved binding energy spectrum for 22 eV primary energy is presented in Fig. 3.47b for M1 magnetization. Since for this (symmetric) scattering kinematics the cross-section for total spin $S = 1$ is zero these spectra reveal the dominance of minority electrons just below the Fermi level. This finding is similar to what was predicted theoretically for spin-polarized (e, 2e) from a 2 ML cobalt film on Cu(001) (Rücker et al. 2005). It was demonstrated that the main features in the (e,2e) spectra for symmetric detection of electrons are related to the majority and minority densities of states at $k_{bx} = 0$ (Rücker et al. 2005). A spin-resolved photoemission experiment on a thin Co film on W(110) with normal electron emission and photon energy 21.22 eV shows a strong dominance of minority states within 1 eV below the Fermi level, which changes to the dominance of majority spin states for $E_b < -1$ eV (Bansmann et al. 1995). The asymmetry of the spectral density function (Fig. 3.47 c) can be extracted from the measured spectra according to (3.72). Figure 3.47c shows such asymmetries measured at various primary energies. All four curves in Fig. 3.47c almost coincide within 1 eV below the Fermi level and show a maximum asymmetry of about −26%. The two curves corresponding to 22 and 27 eV primary energies are close to each other, change sign at about −1.4 eV binding energy and reach a positive maximum of about 10% at −2.2 eV binding energy. The 12 eV curve deviates from these two curves in the binding energy range below −1 eV. This most likely indicates that 12 eV is too low an energy to probe reliably binding energies below −1 eV, because the low-energy final states electrons undergo refraction on the surface potential barrier and are not

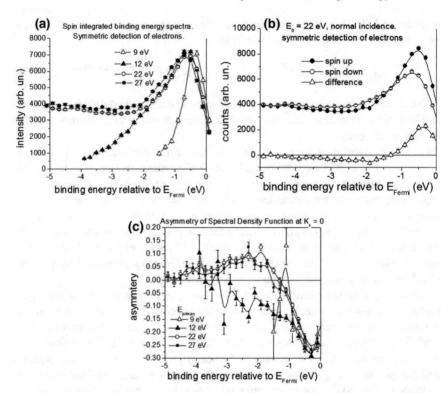

Fig. 3.47 **a** Spin-integrated binding energy spectra with symmetric detection of electrons ($X^{(S=1)}$ $=0$); **b** spin-resolved binding energy spectra with symmetric detection of electrons; **c** asymmetry of spectral density function at $k_x = 0$. (Reprinted from Samarin et al. (2007a), Copyright (2007), with permission from Elsevier)

detected. The multi-step contribution instead becomes a dominant contribution to the spectrum at these energies.

The data presented in Fig. 3.47 correspond to the valence states confined around $k_{bx} = 0$. To visualize the exchange and spin–orbit interactions in the larger range of the valence electron momentum (wave vector) we measured (e,2e) spectra (I^+ and I) for two opposite magnetizations of the sample: $M1$ and $M2$. Then two asymmetries of the k_{bx}-distributions were constructed according to (3.79) with the binding energy confined within 0.5 eV just below the Fermi level: A^{M1} and A^{M2}. Figure 3.48a represents these distributions (as measured, without normalization). These two curves would be mirror symmetric with respect to the x-axis if the spin effect was purely determined by exchange interaction. It is clearly seen that they are not symmetric due to the presence of a spin–orbit interaction. The spin–orbit (A_{so}) and exchange (A_{ex}) contributions to the measured asymmetry are given to leading order, as discussed above, by:

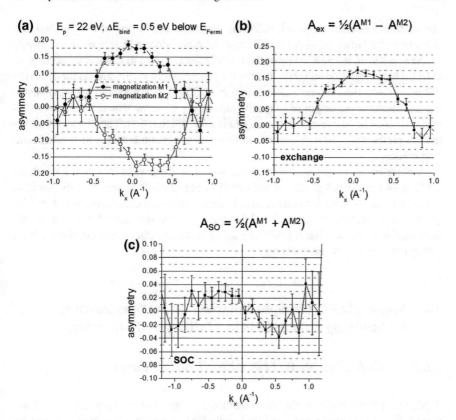

Fig. 3.48 **a** Asymmetries of k_x-distributions measured on 3 ML Co film on W(110) with 22 eV spin-polarized primary electrons; **b** exchange component of the asymmetry; **c** spin-orbit component of the asymmetry. (Reprinted from Samarin et al. (2007a), Copyright (2007), with permission from Elsevier)

$$A_{ex} = 1/2\left(A^{M1} - A^{M2}\right)$$
$$A_{so} = 1/2\left(A^{M1} + A^{M2}\right).$$

The exchange and spin–orbit contributions are shown in Fig. 3.48 (b) and (c), respectively. One can see that the exchange asymmetry is located approximately between $k_{bx} = -0.5\,\text{Å}^{-1}$ and $k_{bx} + 0.5\,\text{Å}^{-1}$ and reaches a maximum of about 17% in the centre of the Brillouin zone.

The spin–orbit contribution exhibits typical "left–right" character and reaches a maximum of -3% at $k_{bx} \approx -0.5\,\text{Å}^{-1}$ and minimum of -3% at $k_{bx} \approx 0.5\,\text{Å}^{-1}$. A similar analysis of the SOI for binding energies farther away from the Fermi level shows that the spin–orbit contribution diminishes for $E_b \leq -1$ eV. This energy location of the SOI in the Co film is compatible with the results of photoelectron spectroscopy (Bansmann et al. 1995; Rampe et al. 1995). The spin–orbit interaction in a ferromagnetic film couples the spin magnetic moment of an electron to the crystal lattice and therefore is often responsible for the magnetic anisotropy of the film, i.e. the existence of a preferred direction of magnetization. Moreover, magnetic anisotropy is correlated with the anisotropy of the orbital moment (Bruno 1989; Ravindran et al. 2001). This correlation suggests that the easy magnetization direction is the direction of the largest orbital moment (Ravindran et al. 2001). It is essential for understanding the microscopic origin of magnetic anisotropy to know which electronic states in the valence band determine the easy axis of magnetization. Therefore, any information on the spin–orbit interaction in thin ferromagnetic layers is of fundamental importance.

A comparison of the spin-orbit components of the asymmetry in the substrate W(110) and in the cobalt film is presented in Fig. 3.49. It is seen that the positive asymmetry in the substrate corresponds to the negative asymmetry of the cobalt film (at the same location in k-space) and this may indicate that the substrate "induces" such spin-orbit interaction (of opposite sign) in the interface region of the Co film, which reduces the interface energy.

3.6 Application of the Spin-Polarized (e,2e) Spectroscopy for Studying Spin Effects in Multilayered Structures

3.6.1 Cobalt Film on W(110) with Ni Buffer Layer

A peculiar interaction between the constituents of multilayered ferromagnetic structures defines the overall magnetic anisotropy of the system and is very sensitive to the crystallinity and the thickness of layers. One striking example is the spin-reorientation transition (SRT) discovered in the Ni(1ML)/Fe(1ML)/Ni(8ML)/W(110) system (Zillgen et al. 1994; Schirmer and Wuttig 1999; Dunn et al. 2005; Sander et al. 1997; Lee et al. 2011), where ML means a monolayer. In this system at room temperature, the easy magnetization axis of the entire system changes from in-plane (8ML Ni/W(110)) to out-of-plane (at 1ML Fe/8ML Ni/W(110)) and back to in-plane (at 1ML Ni/1ML Fe/8ML Ni/W(110)) with successive deposition of additional MLs of Fe and Ni. The microscopic origin of such a complicated behavior of the multilayered system is not yet completely understood. Therefore a study of the geometric structure and spin-related electronic properties of a bi-layer ferromagnetic system such as Co/Ni/W(110) has particular interest. The structural and magnetic properties of a Co film on W(110) (Bansmann et al. 2000; Fritzsche et al. 1995; Knoppe and Bauer 1993a, b; Getzlaff 2001; Garreau et al. 1997) and of Ni films on W(110) (Sander et al. 1997; Farle et al. 1990; Li and Baberschke 1992; Kolaczkiewiczl and Bauer 1984; Sander et al. 1998; Farle and Gradmann 1990; Kämper et al. 1989) are well studied. It was established that a Co film grows on W(110) surface in *hcp* structure with its c axis perpendicular to the surface and easy magnetization axis in the plane of the film along the [1$\bar{1}$0] direction of the substrate. As for a Ni film, it grows in *fcc* structure (above one monolayer (ML)) with the [111] direction perpendicular to the

Fig. 3.49 Comparison of spin-orbit component of asymmetry in the substrate (W(110)) and Co film on W(110)

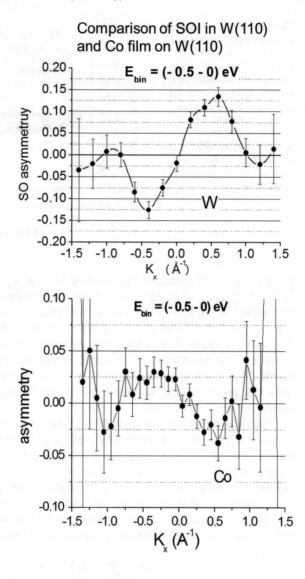

surface. Its easy magnetization axis lies in the plane of the film and is parallel to the [001] direction of the substrate.

A combination of spin-polarized single- and two-electron spectroscopic studies of the bi-layered system with a fixed 3 ML thick Co film and varied thickness of the Ni buffer layer is discussed below. The Ni and Co films were deposited using EFM-3 (OMICRON) evaporators with the deposition rate of 0.5 ML per minute. Ni films were deposited in field-free conditions with the Earth's magnetic field compensated by Helmholz coils down to 10 mG. Co films were deposited as described above (Samarin et al. 2006, 2007a, b). Namely, a weak magnetic field (about 600 mG)

Fig. 3.50 Stoner excitation asymmetries of Co film with and without Ni buffer layer. Solid symbols—3 ML Co film on 2 ML Ni buffer layer on W(110); open circles and squares—3 ML Co film on W(110), magnetizations *M*1 and *M*2, respectively. (Reprinted figure with permission from Samarin et al. (2011a), Copyright (2011) by APS)

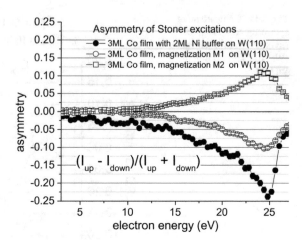

generated by a coil around the vacuum chamber was applied parallel or anti-parallel to the [1$\bar{1}$0] direction along the surface of the tungsten substrate. The direction of magnetization has been defined by the direction of applied magnetic field. This weak magnetic field is sufficient to magnetize a very thin Co film at an initial stage of growth because the Curie temperature of such a film apparently is very low (room temperature of even lower) (Garreau et al. 1997). The thickness of the Co film was always 3ML whereas the thickness of the Ni layer varied from 0 to 9 ML. The quality of prepared films was monitored by Low Energy Electron Diffraction and Auger Electron Spectroscopy.

Prior to the (e,2e) measurements SPEELS measurements were made on deposited ferromagnetic layers. Electron Energy Loss spectra were recorded for spin-up polarization (that we define as polarization vector being parallel to the Y axis) of the primary electron beam and denoted as I^+ (in Fig. 3.37 insertion our definition of "spin-up" coincides with the conventional definition of "spin-up" (Kirschner 1985b PRL) relative to the magnetization M1) and for spin-down polarization of the incident beam (opposite to "spin-up") denoted as I^-. The spin-dependent features of a SPEEL spectrum are visualized by an asymmetry A defined as:

$$A = \left(I^+ - I^-\right)/\left(I^+ + I^-\right).$$

For a ferromagnetic sample the main spin-dependent feature of a SPEEL spectrum is the Stoner excitations asymmetry (Kirschner 1985a, b), which we used as an indication of the magnetization of the film along the quantization axis which is chosen to be the Y axis of the coordinate system (see Fig. 3.1). An example of the Stoner excitation (SE) asymmetries for two opposite magnetizations *M*1 and *M*2 of the Co film on W(110) is shown in Fig. 3.50. The sign of the asymmetry changes when the sign of magnetization changes and this clearly indicates the magnetic origin of this asymmetry.

Ni layers in the thickness range from 1 to 9 ML deposited on the W(110) surface show no SE asymmetry, i.e. no magnetization along the chosen quantization axis (Y axis). When a Co film is deposited on a 2ML Ni buffer layer it shows Stoner excitations asymmetry that is now about two times larger than in the case of no buffer layer (Fig. 3.50). This result indicates the change of the Stoner density of states (DOS). Indeed, the Stoner DOS is a sort of joint density of states with the condition that occupied and unoccupied states have opposite spins and they are separated by a definite momentum transfer (Kirschner 1985a, b PRL; Kirschner and Suga 1986; Kirschner et al. 1984). The sign and the magnitude of the asymmetry of SE are defined by the combination of majority and minority density of states and their energy distribution. The presence of the Ni buffer layer changes this combination due to the interface effects.

For the structural properties of the bi-layer system it seems that a Ni buffer layer serves as a good template for the cobalt film. Comparison of the LEED patterns of the 3 ML Co film on clean W(110) and on a Ni buffer layer shows that Ni template improves the crystallinity of the layer that shows up in much sharper LEED patterns (see Fig. 3.51). Two-electron spectroscopy relies on the fact that the incident electron undergoes a single-step individual electron-electron scattering with the valence electron of the target. The energy conservation law in the (e,2e) reaction implies that $E_b = (E_1 + E_2) - E_o$, where E_b is the binding energy of the valence electron, E_o is the primary electron energy and E_1 and E_2 are the energies of two detected electrons. This energy conservation law only defines the binding energy of the valence electron if the detected electrons result from a single-step electron-electron collision. Earlier experimental results (Artamonov et al. 1997) have demonstrated that the dominant contribution of the single-step process can be ensured for correlated pairs excited from a metal crystal within about 4 eV below the Fermi level. This conclusion has been made on the basis of the measurements at normal and off-normal incidence. It was shown that the correlated electron pairs excited from within this binding energy range "remember" the tangential component of the incident electron at off-normal incidence, i.e. they obey the momentum conservation law for the component parallel to the surface, i.e. $\mathbf{k}_{b\parallel} = \mathbf{k}_{1\parallel} + \mathbf{k}_{2\parallel} - \mathbf{k}_{0\parallel}$, where indices b, 1, 2, 0 denote parallel to the surface components of the valence electron, first and second detected electrons, and incident electron, respectively. This is only possible if the incident electron undergoes "binary" electron-electron collision (i.e. a collision between two electrons) and no additional electron-electron or electron-phonon collisions occur.

Another argument in favor of the individual electron-electron collision in the (e,2e) experiment on metals is the observation of the exchange effects in the scattering of spin-polarized electrons on a ferromagnetic surface (Samarin et al. 2000a PRL). Figure 3.52 shows the spin-resolved binding energy spectra of 3ML Co film on W(110) with 1Ml Ni buffer layer for two opposite magnetizations of the sample. It is seen that the difference between spin-up and spin-down spectra changes sign when the magnetization of the film changes sign. This indicates that the origin of the spin-related effect is the exchange interaction.

Thus, energy and momentum conservation requirements allow the valence electron involved in the collision to be localized in energy-momentum space in contrast

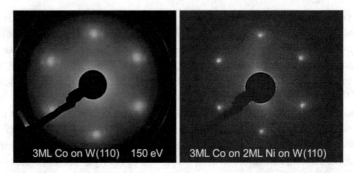

Fig. 3.51   LEED patterns from 3 ML Co film on W(110) without (left) and with (right) Ni buffer layer. (Reprinted figure with permission from Samarin et al. (2011a), Copyright (2011) by APS)

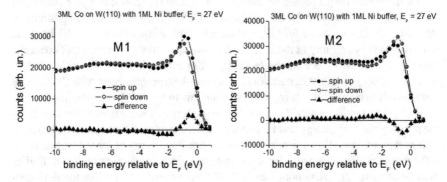

Fig. 3.52   Spin-resolved binding energy spectra for two magnetizations of 3 ML Co film on W(110) with 1 ML Ni buffer layer. (Reprinted figure with permission from Samarin et al. (2011a), Copyright (2011) by APS)

to a single-electron spectroscopy (EELS), where the result of a scattering event is integrated over an unresolved second electron state.

As mentioned previously at the particular kinematical arrangements of (i) normal incidence and (ii) detection of two electrons with equal energies and at equal angles, the asymmetry of the binding energy spectrum represents the asymmetry of the spectral density function of valence electrons in the centre of the Brillouin zone: $S(E, q_x = 0) = (D^+ - D^-)/(D^+ + D^-)$, where D^+ and D^- are density of states for majority and minority electrons at an energy E and a wave vector $q_x = 0$.

In that way the observation provides an access to the spin-dependent electronic structure through the measured (e,2e) spectra. The measured spin asymmetry A of the binding energy spectrum for these pairs (with equal energies and detected at equal angles) would represent spin-asymmetry of the Bloch spectral density function (SDF) in the centre of the Brillouin zone $S(E, k_{\parallel} = 0) = -(1/P)A$, where P is the degree of polarization of the incident beam. Figure 3.53 shows the spin-asymmetry of the SDF of 3ML Co film for two cases: (i) 3ML Co film on W(110) surface and (ii) 3ML

Fig. 3.53 Spin asymmetry of the spectral density function of Co film with and without Ni buffer layer at $k_x = 0$. (Reprinted figure with permission from Samarin et al. (2011a), Copyright (2011) by APS)

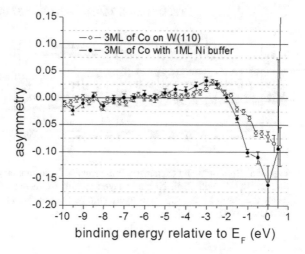

Fig. 3.54 Asymmetry of SDF of the Co/Ni/W(110) system as a function of the Ni atomic layers. (Reprinted figure with permission from Samarin et al. (2011a), Copyright (2011) by APS)

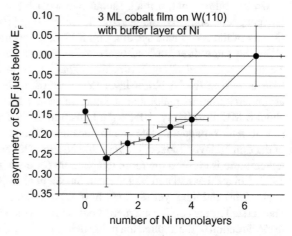

Co film on 1ML Ni buffer layer. It is seen clearly that a Ni buffer layer substantially enhances asymmetry of the SDF just below the Fermi level.

The following measurement was made to show what happens when the Ni film becomes thick and the magnetic moment aligns along the (100) direction in the plane of the W(110) substrate. The asymmetry of the spectral density function of the *Co* film on W(110) with a Ni buffer layer was measured as a function of the thickness of the Ni layer (Fig. 3.54). It is seen that few atomic layers of Ni increases the absolute value of asymmetry of $S(E, k_\parallel = 0)$ compared to cobalt-only case but, starting from the thickness of 2 ML of Ni, the absolute value of asymmetry starts to drop and reaches zero at about 6 ML. The non-vanishing value of the asymmetry $S(E, k_\parallel = 0)$ indicates an imbalance of spin-up and spin-down density of states in the film. A thin (1–4 ML) Ni buffer layer increases the relative minority density of states compared to the Co/W(110) system as shown by the asymmetry of the SDF.

3 Ml of Co film on 2 ML Ni buffer, $E_p = 27$ eV

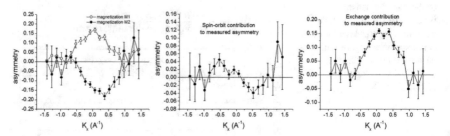

Fig. 3.55 Spin-orbit and exchange components of the asymmetry of K_x-distributions of 3ML Co film on W(110) with 2ML Ni buffer layer; spectra are integrated over 1 eV binding energy just below the Fermi level. (Reprinted figure with permission from Samarin et al. (2011a), Copyright (2011) by APS)

On the other hand, a thick (about 5–6 ML) layer of Ni which is already in a ferromagnetic state and magnetized in-plane along [100] direction of the substrate (X axis of the coordinate system, Fig. 3.17b), drives the Co layer to be magnetized along the same direction and hence to be seen as "nonmagnetic" by the incident electrons polarized along Y direction. That's why in Fig. 3.54 the asymmetry of SDF is zero at the Ni buffer layer thickness of 6 ML. The transition of the SDF asymmetry from its maximum absolute value at 1 ML of Ni buffer layer to zero at about 6 ML most likely reflects the transition of the Ni layer from a paramagnetic state to a ferromagnetic state due to thickness dependence of the Curie temperature of the thin Ni layer. Indeed, for very thin Ni films (in the range of 1–4 ML) the Curie temperature is close to the room temperature (or even below it) (Li and Baberschke 1992; Kolaczkiewiczl and Bauer 1984; Sander et al. 1998; Farle et al. 1990; Kämper et al. 1989) and the layer does not have a preferred magnetization direction and is magnetized by the applied magnetic field along Y axis. The Co film grows magnetized in the direction of the applied magnetic field. After 3 ML Co film is grown the external field is removed and all measurements are done for remanent magnetization.

To test the influence of the W(110) substrate on the spin-orbit interaction in the Co film the (e,2e) spectra were measured for two opposite magnetizations M1 and M2 (both are perpendicular to the scattering plane) of the bi-layered system with various thicknesses of the Ni buffer layer. The exchange and spin-orbit interactions are visualized as before by k_x-distributions of correlated pairs. These distributions represent the number of pairs as a function of the k_x component of the valence electron wave vector at the binding energy of the valence electron within 1 eV below the Fermi level (Fig. 3.55).

It is seen from Fig. 3.55 that a 3ML Co film on 2ML Ni buffer layer exhibits a substantial spin-orbit component in the k_x-distribution asymmetry for binding energy within 1 eV below the Fermi level. The shape and magnitude of the asymmetry are very similar to those observed in a Co film on W(110) substrate without a buffer layer (Samarin et al. 2007a SS). This finding seems to rule out the suggestion about a proximity effect between the W substrate and the Co film that induces the spin-orbit interaction in the Co film. On the other hand, it may happen that a very thin Ni layer still "translates" the spin-orbit interaction into the Co film from the W substrate.

3.6.2 Probing Spin-Related Properties of Magnetic/Nonmagnetic Double Layer on W(110) Using Spin-Polarized (e,2e) Spectroscopy

The spin-orbit interaction, i.e. the interaction of an electron spin with the orbital momentum of the electron, is one of the central features in spintronics (Wolf et al. 2001; Žutě et al. 2004). It is responsible for the Spin Hall effect (Kato et al. 2004; Wunderlich et al. 2005) and, for example, determines properties of topological insulators (Moore 2010). Characterization of those properties follows from measurements of spin-orbit effects and variation of the strength of the spin-orbit interaction (SOI) in a solid structure.

One of the powerful techniques to measure the SOI uses magnetic circular dichroism (Stöhr and Magn 1999), which is element specific and provides quantitative information. The drawback of this technique is that it integrates the results over the Brillouin zone and cannot localize the SOI in the momentum space of the valence band. An alternative technique, which is very surface sensitive and localizes SOI in energy-momentum space is Low-Energy Spin-Polarized Two-Electron Spectroscopy (SPe,2e) (Samarin et al. 2007a, b, 2008, 2011).

As usual, prior to deposition of the films, the substrate W(110) was thoroughly cleaned using the well established routine (Zaker et al. 2010; Cortenraada et al. 2001) including oxygen treatment of the surface followed by high temperature flashes. The 5 monolayers (ML) of Fe film were deposited using the Omicron commercial evaporator EFM-3. On top of an iron film a thin (about 1 ML) Au layer was deposited by evaporating a small piece of gold from v-shaped 0.2 mm diameter tungsten wire by resistive heating. The thickness of the films was estimated by Auger electron spectroscopy and a quartz microbalance. The Auger spectrum of the double layer Au/Fe/W(110) presented in Fig. 3.56 shows clearly the contributions from both Fe and Au layers (Samarin et al. 2015a APL).

Deposition of a gold layer dramatically changes the binding energy spectrum of Fe/W(110) system (Samarin et al. 2015a APL). Figure 3.57 shows comparison of the binding energy spectra of W(110), Fe/W(110) and Au/Fe/W(110). After a Fe film deposition on W(110) the exchange and spin-orbit components of the asymmetry of (SPe,2e) spectra of a pure Fe film on W(110) was checked. As expected there was no detectable spin-orbit component in the measured asymmetry (in contrast to the cobalt film).

When a thin layer of gold was already deposited, as confirmed by Auger electron spectroscopy (Fig. 3.56), the double layer of Au/Fe was still exhibiting ferromagnetic properties. Indeed, the binding energy spectra recorded with spin-up and spin-down

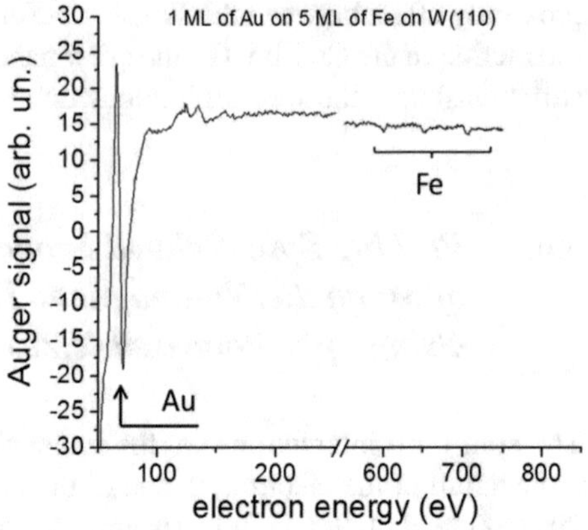

Fig. 3.56 Auger spectrum of the Au/Fe double layer on W(110). (Reprinted from Samarin et al. (2015a) with the permission of AIP Publishing)

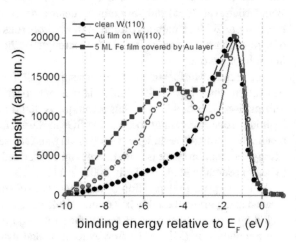

Fig. 3.57 Spin integrated binding energy spectra of W(110), Au film on W(110) and Fe film covered by Au layer. (Reprinted from Samarin et al. (2015a) with the permission of AIP Publishing)

incident electrons show imbalance of spin-up and spin-down states in the valence band of the sample (Fig. 3.58). Comparison of the binding energy asymmetry spectra of pure Fe film and of the Fe film with the gold cap layer shows that the shape of the asymmetry is different in these two cases (Fig. 3.58c). Moreover, the asymmetry spectrum of a double layer of Au/Fe shows a larger value and a larger energy range of asymmetry (Fig. 3.58c).

The change of asymmetry in the binding energy spectrum after an Au layer deposition indicates the modification of the electronic structure of the surface (Fig. 3.57). It also suggests that a thin gold layer may become ferromagnetic. In fact, there are a number of indications that the gold exhibits ferromagnetic properties when its shape indicates nanoparticles (Yamamoto et al. 2004) or thin layers (Kim et al. 2003). However the shape of the binding energy spectrum of a double layer of Au/Fe is different from the spectrum of an Fe film on W(110) and from the spectrum of a Au layer on W(110) (Fig. 3.57).

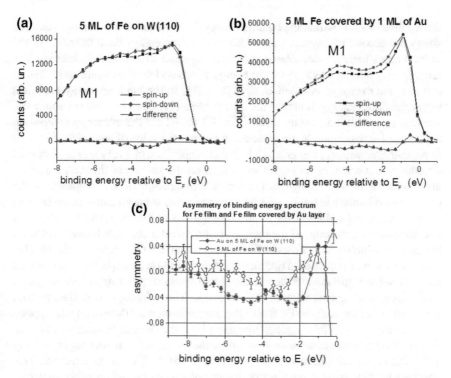

Fig. 3.58 Binding energy spectra for Fe film (**a**) and for Fe film covered by a layer of Au (**b**); asymmetry of binding energy spectra for Fe film and Au/Fe bilayer (**c**). (Reprinted from Samarin et al. (2015a) with the permission of AIP Publishing)

That result confirms the suggestion that the surface electronic structure of the double layer of Au/Fe is modified due to hybridization at the interface of the Au–Fe states. In addition, a strong exchange field of the Fe film can induce a magnetic moment in the Au film, hence an imbalance of spin-up and spin-down states in the valence band. This imbalance shows up as an asymmetry in the binding energy spectrum of the double layer (Fig. 3.58c). The possibility of induced spin polarization in a nonmagnetic layer by an adjacent ferromagnetic layer has been demonstrated (Samant et al. 1994).

An analysis of the K_x-distributions and corresponding asymmetries for the Au/Fe double layer was made using two binding energy bands so that the K_x-distributions were measured with spin-up and spin-down incident electrons. The first band spans from 0 down to -1 eV binding energy, where the difference and asymmetry of the binding energy spectrum are positive (Fig. 3.58b) and the second band from -1 eV down to -2 eV binding energy where the difference and asymmetry are negative. Figure 3.59 shows the exchange and spin-orbit contributions to the asymmetries of the K_x-distributions measured within the above binding energy bands. The left column of Fig. 3.59a and b represents the exchange and spin-orbit asymmetries of the K_x-distributions within the binding energy from -1 to -2 eV. The right column shows corresponding asymmetries within the binding energy band from 0 to -1 eV. In both cases there are contributions of exchange and spin-orbit asymmetries. The latter is very similar in shape in two energy bands and reaches value up to 2%. In contrast, the exchange component of asymmetry, in the binding energy band just below the Fermi energy, is almost flat on average and reaches the value of about 3%, whereas in the binding energy band from -1 to -2 eV the exchange component has a negative maximum in the middle and reaches the value of about 7%. *The most important conclusion of this analysis* is that the deposition of a very thin layer of gold introduces a substantial spin-orbit interaction on the surface of the Au/Fe double-layer, which retains ferromagnetic properties. The signature of ferromagnetism is the exchange asymmetry in the K_x-distributions as well as the asymmetry in the binding energy spectrum. In addition, the electron energy loss spectra of the Au/Fe system for two opposite polarizations of the incident beam reveal a non-zero Stoner excitation asymmetry, which is indicative of a ferromagnetic surface (Kirschner 1985a, b). This asymmetry was also measured two months after the gold layer deposition and showed the same value as just after the deposition and so demonstrated a stable ferromagnetic surface protected against vacuum contamination. These findings are in the line with the results of (Kim et al. 2003) where the magnetic-circular dichroism was applied to reveal that the orbital magnetic moment of the constituent Fe atoms in the bcc Fe–Au alloy film is about twice larger than that of pure Fe. It was suggested that alloy-like regions are formed at the interface of the Au/Fe bilayer structure. This implies a hybridization of the electronic states of Au and Fe and probably enhanced the magnetic moment of Fe (and/or an induced magnetic moment of Au) (Kim et al. 2003).

3.7 Visualizing an Exchange-Correlation Hole Using Spin-Polarized (e,2e) Scattering

3.7.1 Concept of "Exchange-Correlation Hole"

The concept of an exchange—correlation hole (*xc* hole) is important in modern solid state theory (Berakdar 2006; Fulde 1993). It determines the exchange-correlation energy term which is a central part within density functional theory. There are currently intense efforts underway to improve the accuracy of the exchange-correlation term. The importance of this term becomes clear if we recall that it determines essential many-body effects, e.g., magnetism. The main point of this concept is the following. Around each electron, the electronic charge density is reduced such that the charge deficit amounts to exactly one elementary charge. This result is a combination of two effects, namely, the Pauli Exclusion Principle and the repulsive Coulomb interaction (Wigner and Seitz 1933; Slater 1934). The Pauli principle indicates that two electrons with parallel spins cannot be at the same location, while the Coulomb

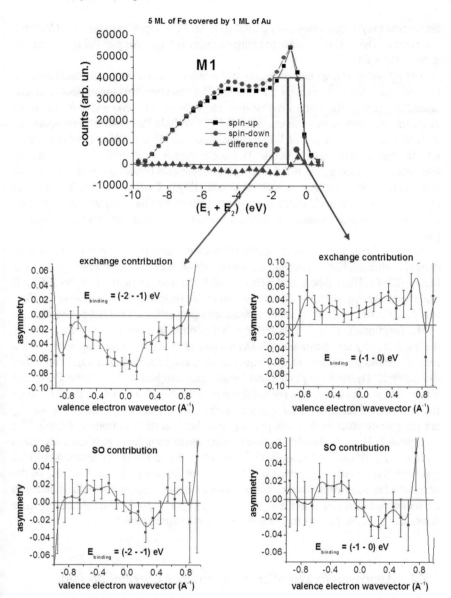

Fig. 3.59 Exchange and spin-orbit contributions to measure asymmetry for two different binding energies. (Reprinted from Samarin et al. (2015a) with the permission of AIP Publishing)

correlation will force electrons to stay apart independently of the spin orientation. The extensions of the exchange part and correlation part of the xc- hole do not have to be necessarily the same. Slater discussed these differences in the pair correlation function (Slater 1934) and indicated for parallel spins both exchange and Coulomb interaction play a role, while antiparallel spin electrons experience only the Coulomb interaction. The region of reduced charge density is larger for parallel spins than for antiparallel spins.

The xc-hole has a spatial extension of the order of a few Angström and amounts, as above mentioned, to exactly the charge of an electron. This means that over distances larger than the size of the *xc*-hole each electron is screened from the other electrons. In other words the electron and the xc-hole look like a 'quasi-particle' with no charge. The work of Wigner and Seitz (1933) and Slater (1934) explains why the approximation of P. Drude, H. Lorenz, and A. Sommerfeld, namely a non-interacting electron gas in metals, can give reasonable results in the description of many electronic properties of metals. Drude (1900), Lorentz (1909), Sommerfeld (1927). However, it is clear that many-body phenomena like magnetism, superconductivity and heavy fermions, to name a few, are beyond the independent electron picture.

A major breakthrough in the description of the ground state properties of condensed matter, including correlation, was the development of the density functional theory (DFT) (Hohenhberg and Kohn 1964; Kohn and Sham 1965; Kohn 1998). It was shown that the ground state energy could be described by a functional of the electron density $n(\mathbf{r})$ and that all other ground state properties are described exactly also by the functionals of $n(\mathbf{r})$. This concept is exploited in a computational scheme by the local density approximation (LDA) leading to the Kohn–Sham equation, which formally looks like a Schrödinger equation of a single particle moving in an effective potential (Hohenhberg and Kohn 1964; Kohn and Sham 1965; Kohn 1998). An important contribution to this potential comes from $\delta E_{xc}/\delta n(\mathbf{r})$, where the functional E_{xc} contains all features of the interacting electronic system. Key quantities for E_{xc} are the pair correlation function $g(\mathbf{r}, \mathbf{r}_0)$ and the *xc*-hole (function) $n_{xc}(\mathbf{r}, \mathbf{r}_0)$. The first quantity is the probability to find an electron at coordinate $\mathbf{r}$, when a second is located at $\mathbf{r}_0$ (Wigner and Seitz 1933; Slater 1934; Fulde 1993). As discussed before, g is essentially constant (normalized to 1) except for small distances $|\mathbf{r} - \mathbf{r}_0|$ of the order of a few angström, where g adopts smaller values. The term *xc*-hole is linked to the pair correlation function via $n_{xc}(\mathbf{r}, \mathbf{r}_0) = n(\mathbf{r}_0)[g(\mathbf{r}, \mathbf{r}_0) - 1]$. It would be therefore desirable to access the key function $n_{xc}(\mathbf{r}, \mathbf{r}_0)$ via experiment.

3.7.2   Approach to Visualize an Exchange-Correlation Hole

Experimental "correlation spectroscopy" techniques involve the momentum-resolved detection of pairs of time-correlated electrons, which emerge from an atom or molecule following the collision of an incoming electron with a bound electron. Electron pair correlation spectroscopy from solid surfaces essentially began more than two decades ago with a prototype reflection mode (e,2e) experiment (Kirschner et al. 1995). Upon impact of primary electrons at low energies (less than about 50 eV), two electrons generated by an incident electron are detected in coincidence by two position-sensitive detectors. Individual energies of both electrons are analyzed by means of a time-of-flight technique. This technique—with various refinements and modifications—has since been demonstrated to be highly surface sensitive and thereby useful for accessing various aspects of the electronic structure of surfaces and for studying electron collision dynamics in the vicinity of surfaces. In particular, surface states were studied experimentally (Samarin et al. 1998 JESRP) and theoretically (Gollisch et al. 2001). Employing spin-polarized primary electrons, the magnetic structure of surfaces has been explored (Samarin et al. 2000a PRL) and spin-orbit coupling effects, which had been predicted theoretically, (Gollisch et al. 1999) experimentally verified (Samarin 2004 PRB). Electron pair correlation spectroscopy from solid surfaces is maturing into a powerful source of direct information on exchange and Coulomb correlation between electrons in condensed matter. The experimental state of the art is highlighted by the recent observation of an exchange-correlation hole in the angular distribution of the two electrons excited from a surface by electron impact (Schumann et al. 2005, 2007a, b) or by photon impact (Hillebrecht et al. 2005; Hattass et al. 2008 (see Chap. 4)).

On the theoretical side, one is confronted mainly with the challenge of calculating Coulomb-correlated two-electron states for a semi-infinite crystalline system, i.e., of solving a non-separable two-electron equation. Formal expressions for the (e,2e) reaction cross section have been derived and presented in detail (Feder et al. 1998 PRB; Berakdar 1998; Berakdar et al. 1999 Solid State Comm). Focusing on valence electrons, the more general Green's function expressions become equivalent to formulae, which essentially consist of matrix elements between an initial two-quasiparticle state $|1, 2\rangle$ and a final two-quasiparticle state $|3, 4\rangle$:

$$I \propto |\langle 3, 4|O|1, 2\rangle|^2 \delta \tag{3.80}$$

According to standard perturbation theory, this form is exact if, for a given perturbation operator W, the initial and final states are the unperturbed ones and O is the transition operator T, which satisfies the equation $T = W + WG_0T$, where G_0 is the unperturbed Green's function. The δ—function in (3.80) ensures pair energy conservation, i.e., $E_{(3,4)} = E_{(1,2)}$. For crystalline surfaces, the δ-function further includes conservation of the two-particle surface-parallel momentum modulo a surface reciprocal lattice vector.

In the case of (e,2e) from solid surfaces, the perturbation W is an effective electron-electron interaction $U(\mathbf{r}, \mathbf{r}_0)$, which is the bare Coulomb interaction if both $\mathbf{r}$ and $\mathbf{r}_0$ are in the vacuum region between surface and detectors, and a screened Coulomb interaction if one or both electrons are inside the solid. The initial state $|1, 2\rangle$ is an antisymmetrized product of the projectile electron state $|1\rangle$—a low-energy electron diffraction (LEED) state with energy E_1, surface-parallel momentum (wave vector) $\mathbf{k}_1^{\parallel}$ and spin σ_1 set asymptotically by the electron gun—and a valence electron state $|2\rangle$ with energy E_2, surface-parallel momentum $\mathbf{k}_2^{\parallel}$ and spin σ_2. Similarly, the unperturbed final state is an antisymmetrized product of two time-reversed LEED states with asymptotic boundary conditions such that an electron with momentum $\mathbf{k}_3$ and spin orientation σ_3 (with respect to a given axis) arrives at one detector and an electron with momentum $\mathbf{k}_4$ arrives at the other detector. Writing the transition operator T as $U(1 + G_0 T)$ and noting that the Moeller operator $1 + G_0 T$ transforms the unperturbed final state into one correlated by U, one can view the matrix element in (3.80) in an equivalent way, which we find more convenient for actual calculations: O is taken as the effective interaction U and $|3, 4\rangle$ as a U-correlated final state with the above boundary conditions.

In order to perform numerical (e,2e) calculations within the above framework, the severe problem is solving a two-electron wave equation with an effective quasiparticle potential (describing the interaction of each electron with the semi-infinite crystalline system) and an effective Coulomb interaction U between the two electrons. While a formal solution is readily provided by the Lippmann-Schwinger equation, the actual calculation is extremely difficult. An approximation, in which the two-body potential U is replaced by repulsive contributions to the one electron potentials, has been employed in (e,2e) calculations (Berakdar 1998 PRB).

On the experimental side there are few experimental arrangements that allow direct observation of the exchange—correlation hole: (i) Observation of angular correlation of electrons; (ii) Energy and momentum correlation of electrons; (iii) Spin-polarized electrons in the (e,2e) experiment enable the contributions of exchange and exchange-correlation hole to be be decoupled from the Coulomb contribution. Consider these approaches briefly (Schumann et al. 2005, 2007a, b NJP, 2010).

To gain information on the electron-electron interaction in a system, focus is given ideally to an electron pair in the system and the probability of finding one electron in some region in momentum space is monitored while changing in a controlled manner the momentum vector of the second electron; i.e., the momentum-space pair correlation function is determined. An appropriate experimental set-up is shown in Fig. 3.60a (Schumann et al. 2005). An incident electron with momentum $\mathbf{k}_0$ and energy E_0 impinges on the sample surface and two time-correlated electrons generated by the incident electron escape the surface with momenta $\mathbf{k}_1$ and $\mathbf{k}_2$, and energies E_1 and E_2, respectively. The electron detector consists of a large channel plate (acceptance angle of about 1 sr) with resistive anode. The central part of the resistive anode is replaced by a separate collector (see Fig. 3.60a) which allows fixing the angular position of the first electron of a pair with the wave vector k_1. The impact position on the resistive anode determines the direction of $\mathbf{k}_2$. In this way the energy and momentum dependence of the electron pair correlation can be mapped out. The experimental energy and momentum resolution are typically 0.5 eV and 0.1 Å^{-1}, respectively. As evident from Fig. 3.60b, the energy, ε, and the wave vector, k, of the valence electron follow from the energy and wave vector conservations, e.g.,

$$\varepsilon = E_0 - E_1 - E_2 - W, \tag{3.81}$$

where W is the energy difference between the vacuum level and the highest occupied level with the energy $\hat{\mu}$ [see Fig. 3.60b].

A crystal of LiF has been chosen as a sample for the following reasons: (i) high coincidence rate compared to metals, (ii) the sample remains clean for a long time if kept at 400 K, (iii) for LiF the energy W is ~14 eV which ensures a good separation

Fig. 3.60 a An electron with momentum $\mathbf{k}_0$ interacts with another electron residing at the top of the valence band of the sample. Two excited electrons with momenta $\mathbf{k}_1$, $\mathbf{k}_2$ and energies E_1 and E_2 are detected in coincidence by a resistive anode and central collector. **b** Energy position of the ejected valence band electron. (Reprinted figure with permission from Schumann et al. (2005), Copyright (2005) by APS)

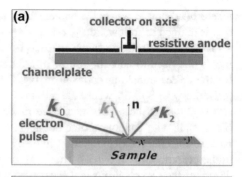

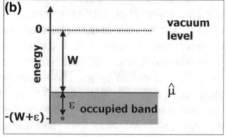

between the elastic peak and electrons ejected from the valence band in the time-of-flight spectrum. In (x) E_0, E_1, E_2 are controllable experimentally and can be chosen such that, according to energy conservation, only one valence band electron is emitted (Schumann et al. 2005).

A key ingredient for the description of the electronic correlation in an N particle system is the reduced (two-particle) density matrix, which is given in terms of the wave function Ψ as:

$$\gamma_2(x_1, x_2, x_1', x_2') = N(N-1)\int \Psi(x_1, x_2, x_3, \ldots, x_N)\Psi^*(x_1', x_2', x_3, \ldots, x_N)dx_3 \ldots dx_N$$

$$(3.82)$$

For fermions this expression dictates that $\gamma_2(x_1, x_2, x_1', x_2') = -\gamma_2(x_2, x_1, x_1', x_2')$. Here $x_j, j = 1, 2, \ldots, N$ stand for spin and spatial coordinates. The two-particle density derives from γ_2 as $\rho_2(x_1, x_2) = \gamma_2(x_1, x_2, x_1, x_2)$. Hence, for fermions for ρ_2 vanishes for $x_1 = x_2 = x$, i.e. $\rho_2(x_1, x_2) = 0$. Even for non-interacting (but overlapping) fermions the antisymmetry of Ψ implies a correlation among the particles that results in the existence of the (Fermi) hole in the two-particle density for $x_1 = x_2$. The Coulomb repulsion between the electrons results in additional contribution to the hole. Usually the hole is quantified by introducing the xc—hole (Fulde 1993):

$$h_{xc}(x_1, x_2) = \frac{\rho_2(x_1, x_2)}{\rho(x_1)} - \rho(x_2),$$

where $\rho(x_1)$ and $\rho(x_2)$ are the one-electron densities. Note that the correlation function is defined as follows: $g(x_1, x_2) \equiv \frac{\rho_2(x_1, x_2)}{\rho(x_1) \cdot \rho(x_2)}$.

To show the relation between $\rho_2(x_1 x_2)$ and the results of the (e,2e) measurements one need to note the following (Schumann et al. 2005): The probability P_{if} for the (e,2e) transition depicted in Fig. 3.60a is given by (Berakdar 2003 book p. 189): $P_{if} = S_{if} \cdot S_{if}^*$ where the S matrix elements are given by $S_{if} = \langle \Psi_{E_f} \mid \Psi_{E_i} \rangle$ and $\Psi_{E_i}(\Psi_{E_f})$ is the normalized wave function describing the system in the initial (final) state with the appropriate boundary conditions. The initial state with energy E_i describes the incident electron interacting with an electron in the valence band in the presences of all other particles in the system. The final state with energy E_f describes the two electrons that escape from the sample. The following approximation (Schumann et al. 2005) that the surrounding medium is not affected while the incident and the valence band electron are interacting during the emission of the two electrons, one can write: $\Psi_{E_i} \approx \psi_{E_i}(x_1, x_2) \cdot \chi(x_3, \ldots, x_N)$. ψ_{E_i} is the electron pair wave function in the initial state with the energy $E_i = E_0 - (\varepsilon + W)$. the surrounding medium is described by χ. The reduced density matrix (3.82) then has the form: $\gamma_2(x_1, x_2, x_1', x_2') \approx 2 \cdot \psi(x_1, x_2)\psi^*(x_1', x_2')$. Furthermore, assuming the emitted electron pair state ψ_{E_f} ($E_f = E_1 + E$) to be described by plane waves, one finds for the measured, spin (σ_j) unresolved probability:

$P_{if} \propto \sum_{\sigma_1, \sigma_2, \sigma_1', \sigma_2'} \tilde{\psi}_{E_i}(\sigma_1 \mathbf{k}_1, \sigma_2 \mathbf{k}_2)\tilde{\psi}_{E_i}^*(\sigma_1' \mathbf{k}_1, \sigma_2' \mathbf{k}_2)$, where $\tilde{\psi}_{E_i}$ is the double Fourier transform of ψ_{E_i}. Hence the measured quantity is the spin-averaged diagonal elements of the reduced density matrix in momentum space, i.e., the spin-averaged momentum-space two-particle density $\rho_2(\mathbf{k}_1, \mathbf{k}_2)$. P_{if} possesses all properties of ρ_2; in particular, P_{if} vanishes for $\mathbf{k}_1 = \mathbf{k}_2$. So, the (e,2e) experiment (Fig. 3.60) allows studying the correlation within the pair consisting of the incident electron coupled to a valence band electron. The indistinguishability of these two electrons contributes to the xc hole through the exchange part.

The measured intensity I of the correlated electron pairs is proportional to the coincident probability P_{if} ($\mathbf{k}_1, \mathbf{k}_2$). The relation between the initial-state correlation and the measured correlation features (in particular, the xc hole) is illustrated below: the pair correlation diminishes when slightly changing the initial state (by changing ε) while keeping the final state (i.e., $\mathbf{k}_1, \mathbf{k}_2$) unaltered. On the other hand, the spectra are hardly affected for the same initial state but different final state energy.

Figure 3.61 shows the energy correlation in the measured electron pair coincidence intensity, $I(E_1, E_2)$ with the direction of $\mathbf{k}_1$ fixed. Because of the low energies entailing a short escape depth, only the electrons from the first few atomic layers of the sample are involved (Morozov et al. 2002; Liscio et al. 2004). Figure 3.61 reveals which electron energies are favoured by the electron-electron interaction at surfaces. We recall that the electron-electron scattering in free space is governed by the form factor of the Coulomb potential that behaves as $(\mathbf{k}_1 - \mathbf{k}_2)^{-2}$. Figure 3.61 indicates, however, a much more complex energetic dependence of the electron-electron correlation function at surfaces. Theory predicts that generally the behaviour shown in Fig. 3.61 is determined by the surface electronic and structural properties (Berakdar et al. 1999 Solid State Commun).

Fig. 3.61 Energy-correlation intensity $I(E_1, E_2)$ (in arbitrary units) with the direction of $\mathbf{k}_1$ fixed. The incident electron has an energy of 30.7 eV and the sample is a LiF(100) surface. (Reprinted figure with permission from Schumann et al. (2005), Copyright (2005) by APS)

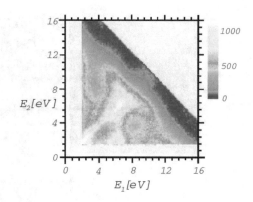

In momentum space, the pair correlation is mapped out as a function of the momentum of one electron (hitting the resistive anode). The momentum of the other electron is fixed by the small central collector (marked by the black dot in Fig. 3.62). The existence of the exchange and correlation-induced hole is evidenced by Fig. 3.62a, b. In free space the electron-pair correlation function is dominated by the factor $2\pi\alpha[exp(2\pi\alpha) - 1]^{-1}$ (Bethe and Salpeter 1977), where $\alpha = 1/|\mathbf{k}_1 - \mathbf{k}_2|$. In a condensed medium, the surrounding charges modify substantially the properties of the electron-electron interaction. While theory cannot provide yet a general expression for $I(E_1, E_2, \mathbf{k}_1, \mathbf{k}_2)$ the experimental findings in Fig. 3.62a, b reveal hole features that are qualitatively different from those known in the free-space or in atomic species (Bethe and Salpeter 1977). Crystal symmetry and the direction of electron momenta determine the 'hole' shape (Fominykh et al. 2002).

If the momentum vector of one electron lies in a crystal high symmetry plane the hole should possess the discrete crystal symmetry. In Fig. 3.62a one cannot clearly resolve the symmetry of the fourfold sample surface due to insufficient statistics. In Fig. 3.62b the alignment of electron momenta with the high symmetry planes is broken upon a 20° rotation of the sample. As evidenced by Fig. 3.62a the hole is shifted so as to surround the fixed electron (black dot), meaning that this hole is, indeed, associated with the fixed electron. A further key issue is the range of the electron-electron interaction in a given sample, which is determined by the size of the hole. In free space the bare electron-electron interaction is of an infinite range and hence $\lim_{|\mathbf{k}_1 - \mathbf{k}_2|\to\infty} I(E_1, E_2, \mathbf{k}_1,\mathbf{k}_2) \to 1/|\mathbf{k}_1 - \mathbf{k}_2|$. When the electron pair is immersed in an electron gas, the electron-electron interaction is screened and the xc hole shrinks. It is only for a diminishing hole (strong screening) that the material can be viewed convincingly as a collection of independent quasiparticles (Fulde 1993).

The energy scale at which the correlation between the selected sample electron and the test charge (incident electron) can be studied is set by $E_0 - (\varepsilon + W)$, which has to be larger than W. A further key finding of this experiment is that the electron-electron interaction, as manifested in the xc hole is very sensitive to de-coherence effects (Schumann et al. 2005). De-coherence sets in when the correlated electron pair scatters inelastically from other surrounding electrons. One can switch off and

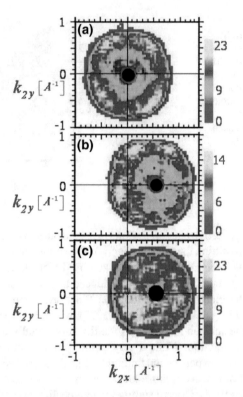

Fig. 3.62 Intensity of the electron correlation function (in arbitrary units) versus the surface momentum components (k_{2x} and k_{2y}) of the electron with energy E_2. The energies E_1 and E_2 are 8 and 9 eV, respectively. The black dot marks the regime where the central collector is. **a** The primary energy is 30.7 eV. **b** As in (**a**) but the sample has been rotated by 20°. **c** As in (**b**), however, the impact energy is increased to 33.7 eV, i.e., the valence electron stems not from top of the valence band but from an energy band with 3 eV width below. Thus, the electron pair may undergo inelastic scattering processes resulting in de-coherence of the electron waves. (Reprinted figure with permission from Schumann et al. (2005), Copyright (2005) by APS)

on such mechanisms by tuning appropriately the electron energies: If E_0, E_1, and E_2 are such that the valence electron involved in the (e,2e) reaction resides at the highest occupied level $\hat{\mu}$, i.e., if $\varepsilon = 0$ [cf. Fig. 3.60b] the electron pair does not inelastically scatter from other particles. Indeed, for such a scattering an energy loss of the electron pair would violate the energy conservation in (3.81). Electron wave coherence is not affected by elastic scattering. On the other hand, as inferred from Fig. 3.60b, keeping the electron pair energies E_1 and E_2 fixed and increasing E_0 we access, in addition to the state with the energy $\varepsilon = E_0 - (E_1 + E_2) - W$, a band of occupied electronic states lying between $\hat{\mu}$ and ε. An electron pair originating from this band has in the solid the excess energy $(E_1 + E_2) + \delta$. Because of inelastic scattering from the occupied levels above ε, the electron pair loses the energy δ and arrives at the detector with the energies E_1 and E_2. These scattering events, whose amount can be tuned by changing δ, randomize the phase of the electron waves and eventually lead to a loss of correlation within the pair. This situation is illustrated in Fig. 3.62c for which the energies E_1, E_2 are kept fixed to be the same as in Fig. 3.62a. When the primary energy E_0 is increased by $\delta = 3$ eV, the de-coherence channel opens, and a complete loss of correlation within the pair is observed. Consequently, the xc hole diminishes; i.e., two incoherent electrons approach each other much closer than in situations where de-coherence is small (absent) and the correlation hole is fully developed [cf. Fig. 3.62a] (Schumann et al. 2005). This finding is of key importance for the potential use of correlated electron pairs in solids for quantum information processing, as discussed in (Saraga et al. 2004). Such applications require coherence within the pair. As found in this study (see also Chap. 2.2) that coherent electron pairs are only those that have the energy $E_0 - W$, i.e., where one electron originates from the top of the valence band. In other words, the electron pair is generated in a single-step electron-electron collision. An implementation of a spin-polarized electron allows the study of spin correlation of magnetic surfaces. In particular, as the electron pair interaction in metals is screened on the scale of few lattice constants the present technique can be used to provide information on the short-range order of the spin projections. Let us consider briefly such approach (Schumann et al. 2010).

It would be interesting to perform experiments aimed to separate the relative size and magnitude of the exchange and correlation contribution to the xc hole. An experimental approach, which is sensitive to the electron-electron interaction, is the electron pair emission from surfaces excited by a sufficiently energetic primary electron. It has been demonstrated above that the concept of the xc hole can be studied by the electron pair emission from surfaces. The xc hole manifests itself through a reduction of the pair emission intensity around the fixed emission direction of one electron, which can be called the depletion zone (Schumann et al. 2007a, Gollisch and Feder 2004 JP; Gollisch et al. 2006 PRB; Schumann et al. 2005; Berakdar et al. 1999). It was shown that the minimum of the momentum distribution correlates with the minimum of the pair correlation function. One may speculate whether spin dependence can be observed in the momentum distribution. For this, one needs to be able to adjust the relative spin orientation of the incoming electron and the target electron. In this way it is possible to "switch off" and "switch on" the contribution due to exchange. Such an experiment requires a spin-polarized electron beam and spin-polarized target (see Fig. 3.63). The generation of a spin-polarized primary beam is an established technique (Pierce and Meier 1976; Samarin et al. 2000a PRL; Morozov et al. 2002). Spin-polarized target electrons are available in ferromagnets, where the overall population of one spin direction (called majority) is larger than for the opposite spin direction (called minority). The imbalance of population of states with opposite spin orientations results in a finite magnetic moment of the sample. The orientation of the majority direction (or the direction of the magnetic moment, which is opposite to the majority spin direction) can be controlled via an external magnetic field or by azimuthal rotation of the sample. The spin-polarized primary beam interacts with the sample along the surface normal. In the coordinate system used the x and y axis are in the surface plane, y being perpendicular to the scattering plane; z axis is perpendicular to the sample surface. Two channelplate-based detectors (labelled

Fig. 3.63 Experimental
set-up for observation of
exchange contribution to
xc-hole

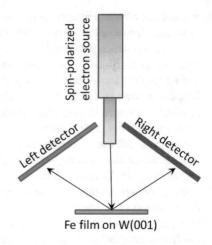

Fe film on W(001)

"left" and "right", see Fig. 3.63) with delay line anodes allow the determination of
the impact positions of the electrons. A coincidence circuit ensures that only one
electron pair per incident electron pulse is detected. A time-of-flight technique was
used for electron energy measurements. From the flight time and the impact position
the kinetic energy and the in-plane components of the momentum for each electron
were determined.

An approximately 20 ML thick epitaxial Fe film on a W(001) surface was used
as a sample. It was magnetized in-plane with the easy axis along the [010] direction
which is parallel to the y-axis. A pulsed magnetic field applied along the y-axis was
used for reversing the magnetization of the sample. The polarization vector of the
incident electron beam is also parallel to the y-axis. The measured (e,2e) spectra
can be grouped into two subsets: (i) for subset I^+ primary spin and majority spin
directions are parallel, (ii) events with antiparallel alignment of primary electron and
majority spin are contained in subset I^-. The reversal of the relative spin orientation
occurred every 5 min. The data acquisition times for both spin alignments are equal
allowing direct comparison of the intensity levels. The target is a single crystal surface
therefore, as mentioned before, the in-plane component of the momentum must be
conserved (modulo a reciprocal surface lattice vector) and this can be written as:

$$\mathbf{k}_\parallel^v + \mathbf{k}_\parallel^p = \mathbf{k}_\parallel^l + \mathbf{k}_\parallel^r = \mathbf{k}_\parallel^{sum} \tag{3.83}$$

On the left side of the equation there is the contribution of the valence electron
$\mathbf{k}_\parallel^v$ and the primary electron $\mathbf{k}_\parallel^p$, while on the right side there is contribution of
the detected electrons $\mathbf{k}_\parallel^l$ ("left") and $\mathbf{k}_\parallel^r$ ("right"). The sum of these two terms is
called sum momentum $\mathbf{k}_\parallel^{sum}$. Since a normal incidence primary beam was used
in this experiment one has $\mathbf{k}_\parallel^p = 0$. It follows that $\mathbf{k}_\parallel^{sum}$ determines the value of the
valence electron wave vector $\mathbf{k}_\parallel^v$. Using energy conservation in the (e,2e) reaction the
energy of the emitted electrons specifies the binding energy of the valence electron.

Therefore, the position of the valence electron within the band structure (in energy-momentum space) is uniquely defined.

To maximize the spin dependence of the observable momentum distribution, one needs energy and momentum conditions, for which target electrons with one spin orientation strongly predominate. Such conditions may be found by means of an ab initio electronic structure calculation. The resulting spin- and layer-resolved densities of states reveal a strong predominance of majority spin for $\mathbf{k}_\parallel^v = 0$ and $E_v = -0.8$ eV, where $\mathbf{k}_\parallel^v$ is valence electron wave vector component parallel to the surface and E_v is the valence electron binding energy relative to the Fermi level. Further, the (e,2e) spectra were calculated with the aims of a quantitative comparison with the experimental spectra and of resolving them with respect to the valence electron spins. For these calculations a formalism was used, which has previously been presented in detail in (Gollisch and Feder 2004 JP; Gollisch et al. 2006 PRB). For the above cases the spin σ of the primary electron parallel and antiparallel to the spin τ of the valence electron, i.e., $\sigma = \tau$ and $\tau = -\sigma = \bar{\sigma}$, the fully spin-resolved (e,2e) reaction cross sections were calculated:

$$I^{\sigma\sigma} \propto |f^{\sigma\sigma} - g^{\sigma\sigma}|^2 \delta \text{ and } I^{\sigma\bar{\sigma}} \propto (|f^{\sigma\bar{\sigma}}|^2 + |g^{\sigma\bar{\sigma}}|^2)\delta, \qquad (3.84)$$

where f is direct scattering amplitude, g is exchange scattering amplitude and δ symbolizes the conservation of energy and surface-parallel momentum. From these partial intensities, summation over the valence electron spins yields the experimentally observable intensities:

$$I^+ = I^{++} + I^{+-} \quad \text{– for incident electron with spin-up;} \qquad (3.85a)$$

$$I^- = I^{-+} + I^{--} \quad \text{– for incident electron with spin-down.} \qquad (3.85b)$$

The aim of this experiment is to disentangle exchange and correlation contribution to exchange-correlation hole, which requires a valence state of high spin polarization. From theory we know that the choice of $\mathbf{k}_\parallel^{sum} = 0$ and a binding energy of 0.8 eV below the Fermi level E_F fulfils this condition while experimentally $|\mathbf{k}_\parallel^{sum}| \leq 0.16$ Å^{-1} allows sufficient counting statistics. Further, the symmetry of the experiment suggests the selection of those coincidence events, where the kinetic energies of both emitted electrons are equal. A primary energy of 25 eV ensures both emitted electrons have a mean energy of 9.7 eV in order to access the selected valence state. For statistical reasons the energy sum of these two electrons needs a window of 1 eV. For 2D momentum distributions of the data, for each coincident event the in-plane components of electron momenta in the left and right detectors are known. According to our coordinate system, k_x^l is always negative while k_x^r is positive. Therefore, a coincidence event has an entry on the left and right half of the plot. In contrast to theory, which covers the full momentum space, the experiment has a limited range. Only momenta which fall inside the area with solid lines as a boundary, can be recorded; see Fig. 3.64a, b. Looking at the experimental intensities I^+ and I^- shown in panels (a) and (b) of Fig. 3.64 one can note that starting at

$|k_x^{l,r}|=0$ the coincidence intensity is zero which is purely instrumental since there is a gap between the detectors. Outside this "blind" region, starting at about $|k_x^{l,r}| = 0.2$ Å^{-1}, an increase of the coincidence intensity for increasing $|k_x|$ values is observed. A maximum is reached at $|k_x^{l,r}| = 0.7$ Å^{-1}. This reduced intensity for small $|k_x^{l,r}|$ values is a manifestation of the xc hole as shown above in experiments and theory (Schumann et al. 2005, Schumann et al. 2007a, b, Fominykh et al. 2002, Gollisch et al. 2006 PRB). Apart from this similarity important differences between I^+ and I^- can be noticed. First, the integrated intensity for I^+ is higher than for I^-. Second, the intensity distribution for I^+ is very different from I^-. Intensities of I^- close to the maximum value are confined to $|k_x^{l,r}|$ values near 0.7 Å^{-1}. For I^+ the intensity levels are close to the maximum value up to $|k^{l,r}_x|$ of 1.1 Å^{-1}, before a decrease can be observed. This is a consequence of the finite angular acceptance of the instrument.

The best comparison of experimental and theoretical data is via line scans through the 2D-momentum distributions for I^+ and I^- in Figs. 3.64e and 3.64f, respectively. The integration range along $|k_y^{l,r}|$ is indicated by the dashed horizontal lines in Fig. 3.64. Both experiment and theory clearly show that the maximum intensity for I^+ is larger than the corresponding maximum for I^-. Further agreement between theory and experiment consists in a larger extension of the depletion zone for I^+. The pair distributions in Fig. 3.64 contain, for each primary electron spin direction, collision events with both a majority- and a minority-spin valence electron according to (3.85a, 3.85b). Further insight is obtained by considering these two events separately. Figure 3.65 shows the calculated four fully spin-resolved intensities $I^{\sigma\sigma}$ and $I^{\sigma\bar{\sigma}}$ [cf. (3.84)]. In Fig. 3.65a the decomposition of I^+ according to (3.85a, 3.85b) is shown. One should note that the main intensity to I^+ comes from the term I^{++} (primary spin-up and valence electron spin-up), whereas the term I^{+-} (primary spin-up and valence electron spin-down) is almost negligible. This is because the chosen valence states are predominantly of a majority type. In other words: the intensity I^+ contains essentially only those collision events, where the spins of the primary and collision partner are parallel. Therefore, exchange and correlation play a role. In a similar way, the intensity I^- is mainly given by the contribution I^{-+} (primary spin-down and valence electron spin-up); see Fig. 3.65b. One can rephrase this by saying that the spin of the collision partner is antiparallel to the spin of the primary and only correlation is important. Again, the origin of this spin selection is that the chosen valence state is essentially of majority type. This intrinsic spin resolution has an important consequence, namely, the possibility to separate exchange and correlation effects between the two outgoing electrons. For the parallel spin case $I^{++} \approx I^+$ both exchange and Coulomb correlation determine the size of the depletion zone, whereas for the antiparallel spin case $I^{-+} \approx I^-$ only the Coulomb correlation plays a role. The size of the depletion zone for I^+ (see Fig. 3.65a) is larger than for I^- (Fig. 3.65b). Therefore, one can say that the size of the exchange depletion zone exceeds the size of the correlation depletion zone.

Experimental results and corresponding pair emission theory provide the first demonstration that it is possible to disentangle exchange and correlation (Schumann

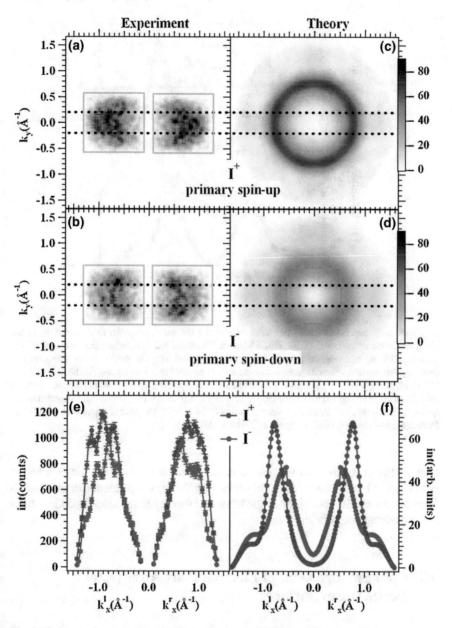

Fig. 3.64 Results for excitation with a primary energy of 25 eV and emission of 9.7 eV electrons. In panels (**a**) and (**c**) the spin polarization of the primary beam and spin direction of the majority electrons are parallel (I^+), while they are antiparallel (I^-) in panels (**b**) and (**d**). Only momenta which fall inside the area, which has the solid lines as boundary can be measured. Panels (**e**) and (**f**) are line scans through the distributions I^+ and I^-) with an integration width determined by the dashed lines in the upper panels. (Reprinted figure with permission from Schumann et al. (2010), Copyright (2010) by APS)

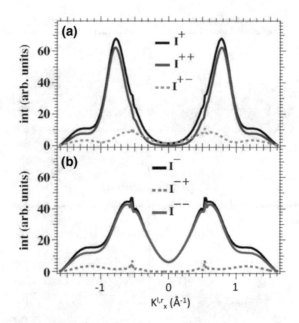

Fig. 3.65 Spin-dependent intensities as calculated for the surface-parallel momentum along the [100] direction (Fig. 3.64f) cf. (3.85a, 3.85b). **a** The solid red line shows the intensity of I^{++} (primary spin-up and valence electron spin-up). The dotted blue line relates to the intensity I^{+-} (primary spin-up and valence electron spin-down). The solid black line represents the sum of $I^{++} + I^{+-} = I^+$. **b** The solid blue line shows the intensity of I^- (primary spin-down and valence electron spin-down). The dotted red line relates to the intensity I^{-+} (primary spin-down and valence electron spin-up). The solid black line represents the sum of $I^{-+} + I^- = I^-$. (Reprinted figure with permission from Schumann et al. (2010), Copyright (2010) by APS)

et al. 2010). It is observed that the depletion zone for exchange is larger than for correlation. Since this zone is closely related to the spin-dependent pair correlation function, the results also apply to the latter and thereby to the spin-dependent parts of the exchange-correlation hole.

3.8 Spin Entanglement of Electron Pairs After (e,2e) Scattering at Surfaces

3.8.1 Definition of Entanglement of a Two-Electron System

In the case of distinguishable particles, a two-particle state is commonly defined as entangled if it cannot be expressed as a single product of two one-particle states. In the case of two identical (spin 1/2) fermions it is however not physically adequate (Ghirardi and Marinatto 2004; Lamata and Leon 2006; Schliemann et al. 2001).

Rather, one should adopt a "physical" definition, according to which a two-particle state is referred to as genuinely entangled if it is not possible to attribute a complete set of properties to both particles individually. Only then does there exist quantum correlations which may violate Bell's inequalities or may be used for teleportation. This statement holds for distinguishable as well as for identical particles. For two electrons, however, it implies that a single Slater determinant, i.e., a single antisymmetrized product of two one-electron states, is not genuinely entangled, whereas it would appear entangled in the sense of not being representable as a single product of one-electron states. A two-electron state $|1, 2\rangle$ consisting of a linear combination of linearly independent Slater determinants with expansion coefficients a_k, which satisfy the normalization condition

$$\sum_k |a_k|^2 = 1,$$

is genuinely entangled if the number of nonzero coefficients a_k (referred to as the Slater number) is greater than one (Feder et al. 2015).

A quantitative measure of the entanglement of $|1, 2\rangle$ state can hence be introduced as a real positive function of these coefficients a_k. Choosing—in analogy with the case of distinguishable particles—the von Neumann entropy $S_N = -tr[\rho_1 \log_2(\rho_1)]$, where ρ_1 is the one-electron reduced density matrix corresponding to the state $|1, 2\rangle$, one obtains [cf., e.g., Ghirardi and Marinatto 2004, (3.12)]:

$$S_N = 1 - \sum_k |a_k|^2 \log_2(|a_k|^2) \tag{3.86}$$

For a single Slater determinant (with only one coefficient $a_1 = 1$), which according to the above definition is not genuinely entangled, one thus obtains $S_N = 1$, while genuinely entangled states are characterized by $S_N > 1$. It is therefore more appropriate to use as a measure of entanglement the modified entropy (Feder et al. 2015):

$$S = S_N - 1 = -\sum_k |a_k|^2 \log_2(|a_k|^2). \tag{3.87}$$

3.8.2 Creation of Entanglement in Electron-Electron Scattering at Surfaces

The actual creation of entangled states by the interaction between particles, especially scattering, is a widely usable mechanism. We consider this possibility following the discussion in (Feder et al. 2015). Let two particles, which are initially far apart and not entangled, move towards each other and after their collision, the two-particle wave function will in general be entangled (Feder et al. 2015). Such entanglement creation by interaction has been theoretically explored for pairs of distinguishable

Fig. 3.66 Scattering of
incident electron by the
valence electron with
opposite spins orientation

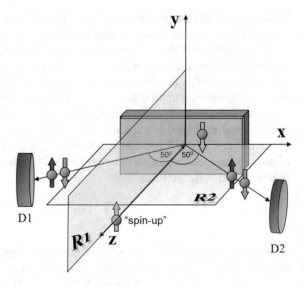

particles and for identical fermions within one-, two-, and three-dimensional models
for various types of interaction, see, e.g., (Wang et al. 2006; Buscemi et al. 2006;
Buscemi et al. 2007; Harshman and Hutton 2008; Harshman and Singh 2008) and
references therein).

A particularly transparent and instructive representative of entangled two-particle
states is the spin-entangled state of two identical fermions, which was proposed in
Bohm's version of the EPR paradox (Bohm 1952) and considered by Bell (Bell
1964) as one of the paradigmatic states, which violate Bell's inequalities and exhibit
nonlocal quantum correlations. This singlet state is hence also referred to as the first
Bell state.

In principle this state, as well as less strongly spin-entangled two-fermion states,
can be created in the scattering of two free identical spin-one-half fermions via
Coulomb interaction [cf. Kessler (1985) and references therein]. A detailed analysis
of the post-collision two-fermion state and its spin entanglement as a function of
the scattering angle have been given (Lamata and Leon 2006). Maximal spin entanglement is obtained in the laboratory system, with one fermion initially at rest, a
scattering angle of 45° and the energy of each fermion is half the energy of the
initially propagating fermion.

Returning now to the specific two-electron states, which are relevant for (e,2e)
reaction that is illustrated in Fig. 3.66.

The primary electron is represented by a low energy electron diffraction (LEED)
state $\left|E_1, \mathbf{k}_1^{\|}, \sigma_1\right\rangle$ with energy E_1, surface parallel momentum $\mathbf{k}_1^{\|}$, and spin orien-
tation σ_1 with respect to an arbitrary but fixed direction. The valence electron state
$\left|E_2, \mathbf{k}_2^{\|}, \sigma_2\right\rangle$ is a bound state with energy E_2, parallel momentum $\mathbf{k}_2^{\|}$, and spin σ_2 . The
two outgoing electrons are described by two time-reversed LEED states $\left|E_3, \mathbf{k}_3^{\|}, \sigma_3\right\rangle$

and $\left|E_4, \mathbf{k}_4^{\parallel}, \sigma_4\right)$ with energies E_3, E_4 and parallel momenta $\mathbf{k}_3^{\parallel}$, $\mathbf{k}_4^{\parallel}$, or equivalently by the corresponding three-dimensional momenta $\mathbf{k}_3$ and $\mathbf{k}_4$, which point into the directions of the two detectors. The spins of the outgoing electrons are labelled by σ_3 and σ_4. With energies and parallel momenta of the primary electron and of the two outgoing electrons fixed by boundary conditions, the respective values for the valence electron are dictated by conservation of energy and of parallel momentum (modulo a parallel reciprocal lattice vector $\mathbf{g}^{\parallel}$):

$$E_1 + E_2 = E_3 + E_4 \quad \text{and} \quad \mathbf{k}_1^{\parallel} + \mathbf{k}_2^{\parallel} = \mathbf{k}_3^{\parallel} + \mathbf{k}_4^{\parallel} + \mathbf{g}^{\parallel} \tag{3.88}$$

To make the following consideration more transparent, we write the above four states as $|n, \sigma_n\rangle = |n\rangle|\sigma_n\rangle$, with $n = 1,4$, where $|n\rangle$ stands for the spatial part $\left|E_n, \mathbf{k}_n^{\parallel}\right\rangle^{\sigma_n}$, which depends on the spin orientation σ_n in the case of a ferromagnetic system.

The initial two-electron state $|I\rangle$ is an antisymmetrized product of the incoming and valence one-electron states:

$$|I\rangle_{\sigma_1\sigma_2} = \frac{1}{\sqrt{2}}(|1, \sigma_1\rangle|2, \sigma_2\rangle - |2, \sigma_2\rangle|1, \sigma_1\rangle) \tag{3.89}$$

First consider the case that the spins of primary and valence electrons are parallel to each other, i.e., $\sigma_1 = \sigma_2 =: \sigma$. In the absence of spin-orbit coupling, the spins of the two outgoing electrons are the same, i.e., $\sigma_3 = \sigma_4 = \sigma$. The only unperturbed normalized final state $|F\rangle$, which is subject to the fixed energy-momentum boundary conditions and allows a nonzero transition amplitude, is then the antisymmetrized product:

$$|F\rangle_{\sigma\sigma}^n = \frac{1}{\sqrt{2}}(|3, \sigma\rangle|4, \sigma\rangle - |4, \sigma\rangle|3, \sigma\rangle) \tag{3.90}$$

The transition amplitude between initial and final states is ${}_{\sigma\sigma}^n\langle F|T|I_{\sigma\sigma}\rangle =: f_{\sigma\sigma} - g_{\sigma\sigma}$, where $f_{\sigma\sigma}$ and $g_{\sigma\sigma}$ are the direct and exchange matrix elements between the spatial parts $|n\rangle = \left|E_n, \mathbf{k}_n^{\parallel}\right\rangle^{\sigma}$:

$$f_{\sigma\sigma} = \langle 3|\langle 4|T|1\rangle|2\rangle \quad \text{and} \quad g_{\sigma\sigma} = \langle 4|\langle 3|T|1\rangle|2\rangle. \tag{3.91}$$

T is the transition operator satisfying the equation: $T = U + UG_0T$, where U is the two-electron interaction potential and G_0 is the unperturbed two-electron Green function. The projected state of two parallel-spin outgoing electrons is thus:

$$|F\rangle_{\sigma\sigma} = (f_{\sigma\sigma} - g_{\sigma\sigma})|F\rangle_{\sigma\sigma}^n \tag{3.92}$$

Then, for the parallel spins of the incident and valence electrons, the observable reaction cross section (intensity) is:

$$I_{\sigma\sigma} = {}_{\sigma\sigma}\langle F|F\rangle_{\sigma\sigma}\delta(E)\delta(\mathbf{k}^{\|}) = |f_{\sigma\sigma} - g_{\sigma\sigma}|^2\delta(E)\delta(\mathbf{k}^{\|}), \tag{3.93}$$

where $\delta(E)$, $\delta(\mathbf{k}^{\|})$, indicate energy and parallel momentum conservation.

In the case of antiparallel spins of the primary and valence electrons, $\sigma_1 =: \sigma$ and $\sigma_2 = -\sigma$, nonvanishing matrix elements exist for two sets of spins of the two outgoing electrons: (a) the direct matrix element $f_{\sigma\bar{\sigma}}$ if $\sigma_3 = \sigma$ and $\sigma_4 = -\sigma$, and (b) the exchange matrix element $g_{\sigma\bar{\sigma}}$ if $\sigma_3 = -\sigma$ and $\sigma_4 = \sigma$, with $f_{\sigma\bar{\sigma}}$ and $g_{\sigma\bar{\sigma}}$ analogous to $f_{\sigma\sigma}$ and $g_{\sigma\sigma}$ in (3.93). The corresponding normalized final states are:

$$|F_d\rangle^n_{\sigma\bar{\sigma}} = \frac{1}{\sqrt{2}}(|3,\sigma\rangle|4,-\sigma\rangle - |4,-\sigma\rangle|3,\sigma\rangle) \quad -\text{direct} \tag{3.94a}$$

$$|F_e\rangle^n_{\sigma\bar{\sigma}} = \frac{1}{\sqrt{2}}(|3,-\sigma\rangle|4,\sigma\rangle - |4,\sigma\rangle|3,-\sigma\rangle) \quad -\text{exchange} \tag{3.94b}$$

These two states are orthogonal to each other. If the outgoing spin set (σ_3,σ_4) is actually fixed the corresponding outgoing two-electron state is then either $f_{\sigma\bar{\sigma}}|F_d\rangle^n_{\sigma\bar{\sigma}}$ or $-g_{\sigma\bar{\sigma}}|F_e\rangle^n_{\sigma\bar{\sigma}}$, and the intensity is either $|f_{\sigma\bar{\sigma}}|^2$ or $|g_{\sigma\bar{\sigma}}|^2$. If the set (σ_3,σ_4) is not fixed, i.e., allowed to be either $(\sigma,-\sigma)$ or $(-\sigma,\sigma)$, the outgoing two-electron state $|F\rangle_{\sigma\bar{\sigma}}$ is the projection of the final state $|F_T\rangle = T|I\rangle$ on to the subspace spanned by the two orthonormal states $|F_d\rangle^n_{\sigma\bar{\sigma}}$ and $|F_e\rangle^n_{\sigma\bar{\sigma}}$:

$$|F\rangle_{\sigma\bar{\sigma}} := f_{\sigma\bar{\sigma}}|F_d\rangle^n_{\sigma\bar{\sigma}} - g_{\sigma\bar{\sigma}}|F_e\rangle^n_{\sigma\bar{\sigma}} \tag{3.95}$$

The reaction cross section for the case of antiparallel spins of the incident electron and the valence electron is then:

$$I_{\sigma\bar{\sigma}} = {}_{\sigma\bar{\sigma}}\langle F|F\rangle_{\sigma\bar{\sigma}}\delta(E)\delta(\mathbf{k}^{\|}) = \left(|f_{\sigma\bar{\sigma}}|^2 + |g_{\sigma\bar{\sigma}}|^2\right)\delta(E)\delta(\mathbf{k}^{\|}) \tag{3.96}$$

Now, one can note that the initial state $|I\rangle$ is just a single Slater determinant. It is therefore not entangled, and according to equation $S = S_N = 1 = -\sum_k |a_k|^2 \cdot \log_2(|a_k|^2)$ it has entropy $S=0$. The same holds for the final state $|F\rangle_{\sigma\sigma}$ [cf. (3.92)] obtained in the case of parallel spins of the two outgoing electrons.

In contrast, the antiparallel-spin final two-electron state $|F\rangle_{\sigma\bar{\sigma}}$ [cf. (3.95)] is a linear combination of two antisymmetrized products of two one-electron states. Since in general both coefficients $f_{\sigma\bar{\sigma}}$ and $g_{\sigma\bar{\sigma}}$ are nonzero, it is genuinely entangled (with Slater number 2). In order to calculate its entropy S, we normalize it, replacing $f_{\sigma\bar{\sigma}}$ and $g_{\sigma\bar{\sigma}}$ by

$$\bar{\bar{f}} = f_{\sigma\bar{\sigma}}/\sqrt{|f_{\sigma\bar{\sigma}}|^2 + |g_{\sigma\bar{\sigma}}|^2} \tag{3.97a}$$

and

$$\bar{\bar{g}} = g_{\sigma\bar{\sigma}}/\sqrt{|f_{\sigma\bar{\sigma}}|^2 + |g_{\sigma\bar{\sigma}}|^2} \tag{3.97b}$$

such that $\left|\bar{\bar{f}}\right|^2 + \left|\bar{\bar{g}}\right|^2 = 1$. According to (3.14) we then obtain

$$S = -\left|\bar{\bar{f}}\right|^2 \log_2\left|\bar{\bar{f}}\right|^2 - \left|\bar{\bar{g}}\right|^2 \log_2\left|\bar{\bar{g}}\right|^2 \tag{3.98}$$

Because of the absolute squares, S can be expressed in terms of the direct and exchange intensities $|f_{\sigma\bar{\sigma}}|^2$ and $|g_{\sigma\bar{\sigma}}|^2$. If the direct matrix element $f_{\sigma\bar{\sigma}}$ has some finite value and the exchange matrix element $g_{\sigma\bar{\sigma}}$ is zero, or vice versa, it is obvious from (3.97a, 3.97b) and (3.98) that the entropy S is zero, and according to (3.94a, 3.94b) and (3.95) the two-electron state $|F\rangle_{\sigma\bar{\sigma}}$ consists of a single Slater determinant. The maximal entanglement, which is possible for $|F\rangle_{\sigma\bar{\sigma}}$, is attained if $|f_{\sigma\bar{\sigma}}|^2 = |g_{\sigma\bar{\sigma}}|^2$. One then has $S = \log_2 2 = 1$.

If $f_{\sigma\bar{\sigma}} = g_{\sigma\bar{\sigma}}$, $|F\rangle_{\sigma\bar{\sigma}}$ can be rewritten in the form:

$$|F\rangle_{\sigma\bar{\sigma}} = f_{\sigma\bar{\sigma}} \frac{1}{\sqrt{2}}(|3\rangle|4\rangle + |4\rangle|3\rangle)(|\sigma\rangle|-\sigma\rangle - |-\sigma\rangle|\sigma\rangle) \tag{3.99}$$

i.e., as a product of a symmetric spatial part and an antisymmetric spin (singlet) part. This is in fact the form of the paradigmatic two-electron state, which was employed in Bohm's version (Bohm 1952) of the Einstein-Podolsky-Rosen paradox (Einstein et al. 1935) and which clearly violates a Bell inequality [cf., e.g., (Ghirardi and Marinatto 2004; Clauser and Shimony 1978) and references therein].

In general, $f_{\sigma\bar{\sigma}}$ and $g_{\sigma\bar{\sigma}}$ depend on the specific surface system and the energies and surface-parallel momenta of the primary electron and of the outgoing electrons. They have to be calculated numerically and yield S somewhere between zero and the maximal value 1.

However, there is a special configuration which always leads to the maximal entanglement. Consider the coplanar symmetric geometry with normal incidence of the primary electron on to the surface (Fig. 3.66) and equal energies ($E_3 = E_4$) and polar angles of the two outgoing electrons. If the reaction plane is perpendicular to a mirror plane normal to the surface or if the surface normal is a twofold rotation symmetry axis, symmetry entails $|f_{\sigma\bar{\sigma}}|^2 = |g_{\sigma\bar{\sigma}}|^2$ and hence from (3.97) and (3.98) $S = \log_2 2 = 1$, which is the maximal entanglement possible for the two-electron state $|F\rangle_{\sigma\bar{\sigma}}$ (3.91).

The occurrence of this maximal entanglement for antiparallel spins is universal in two respects. First, for a given crystalline surface system, it does—for any chosen primary energy E_1 and subject to energy conservation - not depend on the values of the outgoing electron energies $E_3 = E_4$ and not on the surface-parallel momenta $k_3^{\parallel} = -k_4^{\parallel}$, i.e., not on the polar angle $\vartheta_3 = \vartheta_4$ between the emission directions and the surface normal. Secondly, it does not even depend on the choice of a specific surface system (Feder et al. 2015).

The finding of maximal entanglement for all values of the polar angle $\vartheta_3 = \vartheta_4$ and all values of the outgoing electron energies $E_3 = E_4$ is in contrast to the situation in the scattering of two free electrons via Coulomb interaction (Lamata and Leon 2006), where—in the laboratory system with one electron initially at rest—maximal entanglement is generated only if—for a given primary energy E_1—each of the two outgoing electrons has energy $\frac{E_1}{2}$. This in turn occurs only for the special value 45° of the scattering angle.

So far, the cases of parallel and of antiparallel spins have been considered separately. This is realistic for (e,2e) from surfaces if - for valence electron energy E_2 and parallel momentum $\mathbf{k}_2^\parallel$ as determined by the conservation laws (3.88)—only valence electrons with one definite spin orientation contribute to the reaction cross section, which is possible first for ferromagnetic surface systems and second by virtue of (e,2e) selection rules (Feder et al. 2001), even for nonmagnetic systems.

If valence electrons of both spin orientations contribute, the final two-electron state is a mixed state ρ_F described by the sum of the statistical operators corresponding to the two pure states $|F\rangle_{\sigma\sigma}$ (3.92) and $|F\rangle_{\sigma\bar{\sigma}}$ (3.95):

$$\rho_F := |F\rangle_{\sigma\sigma}\langle F|_{\sigma\sigma} + |F\rangle_{\sigma\bar{\sigma}}\langle F|_{\sigma\bar{\sigma}} \tag{3.100}$$

An appropriate measure of the entanglement is then the "entropy of formation" $\bar{S}$ of ρ_F [cf., e.g., (Bennett et al. 1996)]. Since the entropy is zero for the case of parallel spins, we obtain:

$$\bar{S} = SI_{\sigma\bar{\sigma}}/(I_{\sigma\bar{\sigma}} + I_{\sigma\sigma}), \tag{3.101}$$

where S is the entropy for the antiparallel-spin case (3.98) and $I_{\sigma\bar{\sigma}}$ and $I_{\sigma\sigma}$ are the intensities for antiparallel and parallel spins (3.93 and 3.96), respectively. $\bar{S}$ is thus generally smaller than S due to the contribution of parallel-spin electrons.

The value of entanglement S of a pair of electrons resulting from the (e,2e) scattering from a surface can be calculated numerically for various combinations of energies E_3 and E_4 and polar angles ϑ_3 and ϑ_4. The map of S calculated for various combinations of these parameters and comparison with the map of cross section of the (e,2e) reaction can be found in (Feder et al. 2015).

3.8.3 Polarization of a Single Electron and Entanglement of the Pair

Electron systems of two electrons provide a unique possibility to detect spin-entangled states. The two-particle system corresponds to the totally entangled singlet state if the spatial part of the wave function is symmetric relative to the coordinate exchange and the spin part of the wave function is anti-symmetric (Bouwmeester et al. 2000). This is singlet state and the polarization vector of the pair equals zero. When any polarization projection of the scattered electron is being measured then collapse of the two particle wave function occurs and the polarization vector of the detected electron of the entangled pair takes the value up or down in the measurement basis. Detecting of the second scattered electron of the pair by another detector takes the opposite value (down or up in the measurement basis).

The pair of electrons before scattering forms a non-entangled or a separable state. A spin-entangled electron pair may be obtained, for example, as a result of electron-electron scattering, the entanglement degree being dependent on the scattering symmetry and the scattering angle θ. Figure 3.67a represents the scattering symmetry which forms the totally entangled electron pair (left panel). Figure 3.67b represents the scattering conditions which form the partially entangled (or partially separable) electron pair. After the scattering of the two electrons e_1 and e_2 with anti-parallel polarization (red arrows), the pair of two scattered electrons appears and measurements are made of the polarization of one of the pair (the right one). If one chooses the "symmetric" scattering conditions with the two scattered electrons having equal energy and the scattering angle of $\theta_{lab} = 45°$ (in the laboratory frame), only electrons of totally entangled pairs will be detected. Since the spin state of the individual electron in the entangled pairs is not defined, the expectation value of the polarization vector of these electrons would be zero. The "degree" of entanglement of the pair of scattered electrons may be changed by changing the kinematics of scattering. If, for example, one chooses then to detect electrons at a very small scattering angle (Fig. 3.67 right panel), then the "separability" would be close to unity and the electron-electron scattering may result in formation of both entangled and non-entangled pairs. If the entangled pairs are not created specially (for example, by choosing the scattering kinematics) then the polarization vector of the electrons measured by one of the detectors has a non-zero value. It follows that the fact of measurement of zero polarization in the scattering of initially polarized electrons forming singlet pairs can be considered as evidence of the presence of entangled states. Thus, for symmetric scattering a zero-polarization vector of scattered electrons may be considered as a signature of entanglement.

Generally, the two-particle wave function can be presented as a superposition of separate parts, which consists either of a symmetric spin wave function and an anti-symmetric space wave function or of an anti-symmetric spin wave function and a symmetric space wave function. But, in the case of the electron-electron scattering at the scattering angle $\theta_{lab} = 45°$ in the coordinate system where one of the electrons is at rest, the conditions are realized in which the two particle wave function has a symmetric space part and an anti-symmetric spin part (Landau and Lifshitz 1991). For two-particle states the anti-symmetric spin wave function corresponds to a singlet state or to one of Bell maximum entangled spin states (Horodecki et al. 2009). The degree of the interaction in electron-electron scattering is described by the transferred momentum or by the scattering angle. For the pair of electrons detected at given spatial locations r_1 and r_2 a change of scattering angle to obtain "non-symmetric" scattering results in the spin wave function becoming a superposition of singlet and triplet states and the interacting pair is no more in the maximum entangled state.

Various approaches have been used for quantitative description of quantum correlations or the degree of the state entanglement. In one of the first publications (Bennet et al. 1996) the entanglement degree is determined by the state entropy in terms of von Neumann entropy $E(\rho_1) = E(\rho_2)$, where

Fig. 3.67 Two different kinematics allow selection of electrons from totally or partly entangled pairs. The left panel represents the "symmetry scattering" conditions. The right panel represents the common scattering conditions. Also e_1 and e_2 are the primary and the target electrons with anti-parallel polarization (red arrows). The abbreviation "A+MD" means an energy analyser plus a Mott detector. (Reprinted from Artamonov et al. (2015), Copyright (2015), with permission from Elsevier)

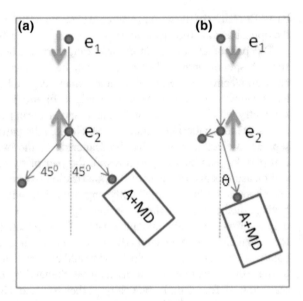

$E(\rho_{1(2)} = -Tr[\rho_{1(2)}\log_2(\rho_{1(2)})])$ $E(\rho_{1(2)} = -Tr[\rho_{1(2)} * Log_2(\rho_{1(2)})])$ and $\rho_{1(2)}$ is the reduced density matrix. The entropy is equal to a logarithm of the number of states which the given physical subsystem can assume (Kadomtsev 1994). In the present case the logarithm is taken to the base 2, which corresponds to the unit of information. To take a logarithm of a matrix it is necessary to diagonalize the matrix, i.e. the logarithm is taken of eigenvalues. For entangled state the subsystems have non-zero entropy as one can see considering the reduced density matrices ρ_1 and ρ_2 while the entropy of a pure state of a composite quantum system is equal to zero. Below, for convenience of comparison of the angular dependence of polarization and entropy, we introduce a linear function of entropy $S(\theta, \Omega) = 1 - E(\theta, \Omega)$, which as antonym to entanglement, characterizes their separability and the possibility to factorize the states. Then $S = 0$ corresponds to non-separable, totally entangled system and $S = 1$ corresponds to the case of totally separable system. The separability of reduced density matrix $\rho_{1(2)}$ is $S(\rho_{1(2)}) = 1 + Tr[\rho_{1(2)} * \log_2(\rho_{1(2)})]$.

Figure 3.68a presents the "first" scattered electron polarization after averaging over all parameters of the "second" one for all possible angles Ω and φ. In the scattering angle region $\theta \approx \pi/2$ the polarization decreases and goes to zero despite the fact that the polarization of electrons before scattering was equal to one. The cause of this is that at this scattering angle there exists the only singlet pair state which corresponds to the maximum entanglement of states (Bouwmeester et al. 2000; Blum 1996). The entanglement of the spin states (or the quantum correlation) manifests itself by impossibility to describe the system characteristics (in the present case spin) consistently and separately for both electrons, i.e. the system is not separable. The pair of interacting electrons in entangled singlet state has total spin equal to zero, but the information about the individual electron spin is lost. When one of the

Fig. 3.68 The polarization P and separability S of the scattered electron beam as a function of the scattering angle θ and relative angle Ω between the polarization vector of the primary electron and that of the electron of the magnetized target at the azimuthal angle $\varphi = 0$. **a** $P(\theta, \Omega)$, **b** $S(\theta, \Omega)$, **c** cross sections of Fig. 3.68a, b at $\Omega = 160°$, $P(\theta)$—the red (long dashed) curve, $S(\theta)$—the blue (solid) curve, S_{ap}—the blue (dashed) curve. (Reprinted from Artamonov et al. (2015), Copyright (2015), with permission from Elsevier)

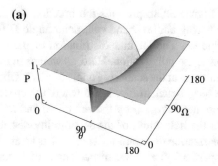

(a)

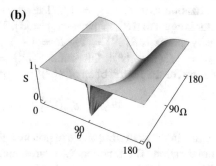

(b)

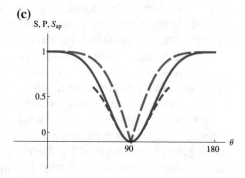

(c)

electrons of the pair is registered its spin projection takes one of two random values corresponding to up or down directions. Because of the strong quantum correlation the spin projection of the second electron assumes the opposite value, which may be verified by registration of the spin projection of the second electron of the pair. Quantum correlations occur regardless of the distance in space between the electrons. There is small probability for investigating experimentally in the foreseeable future the correlation of the spin projections of both scattered electrons. So we consider mathematical expectation of projector, that is, polarization of states.

Figure 3.68b presents the function $S(\theta,\Omega)$ for the whole possible range of the scattering angles θ and spin polar angle Ω for the zero value of the spin azimuthal angle φ. The angular distribution of the separability is, on the whole, similar to that of the polarization. Here, as well as in Fig. 3.68a, the function goes down to the minimum at the scattering angle $\theta = \pi/2$. Differences in the shape of the distributions are seen better in Fig. 3.68c where the cross sections of Fig. 3.68a at $\Omega = 160°$ are presented: $P(\theta,\Omega)$ by the red curve and $S(\theta,\Omega)$ by the blue curve. It can be seen that the half width of the separability distribution significantly exceeds that of the polarization distribution. There exists an analytical relationship between the entropy (separability) and the mean length of a polarisation vector which will be shown below. After diagonalization of the reduced density matrix $\rho_{1(2)}$ of the two particle system one gets the diagonalized matrix ρ^D and two eigenvectors correspondingly. One can define the expectation to find the spin directed parallel (antiparrelel) to the quantization axis as $w_{1(2)} = 1/2(1 \pm d)$, where the mean length of a polarization vector is equal to $|\langle P \rangle| = |w_1 - w_2| = |d|$. The separability function of diagonalized density matrix is $S(\rho^D) = 1 - E = 1 + \sum_n w_n \log_2 w_n$, where E is the entropy and $w_{1(2)} = 1/2(1 \pm |\langle P \rangle|)$. Of course $S(\rho^D) = S(\rho_{1(2)})$. It is possible to reduce this equation for the region of small $|\langle P \rangle|$ by expanding the $log_2 p_n$. It gives:

$$S_{ap}(\rho^D) \approx |\langle P \rangle|^2 \log_2 e/2. \qquad (3.102)$$

Thus, in the scattering angle region near $\theta \approx \pi/2$ separability is a square function of polarization. The function S_{ap} is presented in Fig. 3.68c by the dashed line.

The presented results show that the polarization of electrons scattered due to the electron-electron interaction of two polarized electron ensembles may go to zero only in those cases when the spin state of two interacting electrons becomes entangled. This, in its own turn, occurs in "symmetric scattering", i.e. the polarization of scattered electrons always goes to zero at the scattering angle $\theta \approx \pi/2$. In this case the evaluation of the electron system entropy gives the maximum value or the system separability assumes the minimum value, i.e. the electron system becomes maximum entangled and non-separable. These conclusions are valid only in the frame of conditions given above, i.e. for the idealized case of "pure" electron-electron scattering, which could be understood as the scattering of free primary electrons with a given polarization vector on free target electrons with arbitrary polarization. The scattered electron depolarization may also appear due to other reasons, for example, as a result of additional interaction of the detected electrons (with electrons, phonons, atoms) leading to randomization of the spin direction. One would like to add that in the case of the interaction of electrons with the oppositely directed spins the scattered electron polarization not only decreases down to zero at $\theta \approx \pi/2$, but also changes sign in the $\theta > \pi/2$.

$$\theta > \pi/2.$$

Fig. 3.69 Sketch of the (e,2e) set-up with polarized incident electron beam and polarisation measurement in one arm of the spectrometer. (Reprinted figure with permission from Vasilyev et al. (2017), Copyright (2017) by APS)

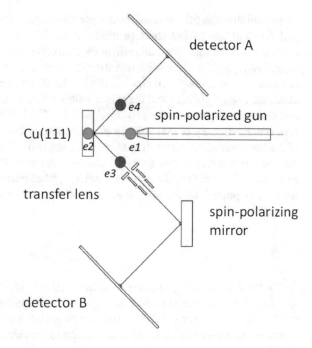

3.8.4 Experimental Demonstration of Spin Entanglement in (e,2e) Scattering

The signature of entanglement of correlated electron pairs emitted from a solid surface under electron impact predicted theoretically in (Feder et al. 2015), was qualitatively demonstrated experimentally using spin-polarized incident electrons and Cu(111) as a sample (Vasilyev et al. 2017).

The authors demonstrated that the von Neumann entropy, which they used to quantify the entanglement, is to be closely related to the spin polarization of the emitted electrons. Therefore, the measurement of the spin polarization of one of the emitted electrons facilitated an experimental study of the entanglement. Experimental set-up that allowed such measurements is shown in Fig. 3.69. In an (e,2e) process, a primary electron e_1 hits a sample and interacts with a valence electron labelled e_2. This leads to the emission of a pair of electrons e_3 and e_4. The time-of-flight coincidence spectrometer with a pulsed spin-polarized electron gun was used for the measurements (Schumann et al. 2010). The spin polarization was around 60% with polarization vector perpendicular to the paper plane. The polarization was switched between the states "+" and "−" by reversal of the helicity of the light source. The primary beam propagates along the sample normal. The arrangement of the detector B in Fig. 3.69 was such that before electrons e_3 can reach this detector, a transfer lens focusses them onto a spin-polarizing mirror (Kirschner et al. 2013; Vasilyev et al. 2015; Tusche et al. 2015).

Essentially, this mirror consists of a pseudomorphic monolayer of Au on Ir(100) and the (0,0) beam of elastic reflection was used. Due to the spin-orbit interaction of electrons scattering from this mirror the reflection coefficient of the mirror for the electrons depends on their spins orientation. Thus it may serve as a spin detector. The detectors A and B consist of channel plates with delay-line anodes. The information about the impact position and flight times allows determination of the kinetic energy and emission angles of the electrons e_3 and e_4. The key quantity to measure is the spin-polarization P_3 of electrons e_3 as a function of the energy difference $E_3 - E_4$ of the two emitted electrons. If we denote the intensity I^+ and I^- for the coincidence measurement with primary spin "+" and "−", the term P_{gun} takes into account the spin polarization of the primary beam, and S_{spin} takes into account the spin sensitivity (analysing power) of the spin analysing mirror, then the spin polarization P_3 is determined as:

$$P_3 = \frac{1}{P_{gun}} \frac{1}{S_{spin}} \frac{I^+ - I^-}{I^+ + I^-}. \tag{3.103}$$

For the antiparallel-spin case (when spin of the incident electron is antiparallel to the spin of the valence electron taking part in the collision) with maximal entanglement at the surface, the value $P_3 = 0$ would be obtained at the detector. $P_3 = 0$ is thus a necessary, but by itself not sufficient condition for entanglement at a macroscopic distance.

In the calculated (e,2e) spectrum from Cu(111) (Fig. 3.70) the dominant feature for pairs excited from the vicinity of the Fermi level (about 0.4 eV below E_F) is the contribution from the sp-like Shockley surface state [cf. Courths et al. 2001 and references therein]. For studying entanglement, this state has the important virtue of even symmetry with respect to mirror planes normal to the surface. If such a mirror plane exists normal to the chosen (e,2e) reaction plane, selection rules (Feder and Gollisch 2001 Solid State Comm) dictate that for $E_3 = E_4$ the parallel-spin intensity $\left(|f - g|^2\right)$ is zero and cannot degrade entropy and spin polarization. As the geometry sketch in the first row of Fig. 3.69 (right side) indicates, this is the case for $\phi = 0°$. Consequently, at the bottom of the Shockley state arch (where $E_3 = E_4$), the entropy (entanglement) has the maximal value one. The intensity difference shown in Fig. 3.70c and the spin polarization P_3 are zero. Sizeable entropy values are seen to extend out to about $E_3 - E_4 = \pm 2$ eV. Going away from the centre, the intensity difference and P_3 become finite, with opposite sign. For azimuthal angles $\phi = 30°$ (second row of Fig. 3.70 and $\phi = 90°$ (third row)), the plane normal to the reaction plane is not a mirror plane. Consequently, collisions between parallel-spin electrons will generally contribute a nonzero intensity term $|f - g|^2$. While the total intensity (Fig. 3.70d, g) is only mildly affected, entropy and intensity difference are changed drastically. In both Fig. 3.70e, h, S is seen to be strongly reduced in the central region of the Shockley arch, and the intensity difference (in Fig. 3.70f, i) is now large, with opposite signs for $\phi = 30°$ and $\phi = 90°$. For the spin polarization P_3, which is the intensity difference divided by the total intensity, this means that—in

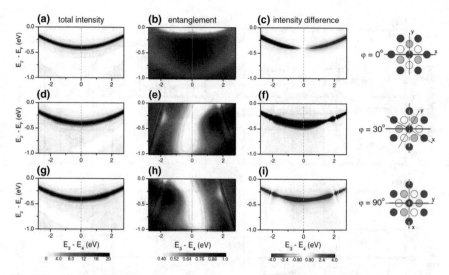

Fig. 3.70 Calculated intensity of the (e,2e) spectrum, entanglement, intensity difference between I^+ and I^- spectra and mirror symmetry of the sample at various azimuthal positions. (Reprinted figure with permission from Vasilyev et al. (2017), Copyright (2017) by APS)

contrast to crossing zero at $E_3 - E_4 = 0$ in the case $\phi = 0°$—it has sizable positive (negative) values for $\phi = 30°$ ($\phi = 90°$).

As mentioned above, the entanglement of the electron pairs is very closely associated with the spin polarization P_3 of the electrons leaving the Cu surface in one direction. Since the sign of P_3 is reversed by reversing the sign σ_1 of the primary electron spin, P_3 can be measured as the asymmetry after spin-dependent reflection at the spin-polarizing mirror (1) (see Fig. 3.69). Experimental spin polarization data are presented, together with theoretical results, in Fig. 3.71. Due to experimental energy resolution, these data have been collected over an energy interval E_2 of 1 eV below the Fermi energy, which comprises the Shockley surface state, and a number of intervals of $(E_3 - E_4)$. The corresponding theoretical results are the average polarization curves $P_3(E_3 - E_4)$, which were calculated from:

$$P_3 = \sigma_1 \frac{(|f|^2 - |g|^2 + |f - g|^2)}{(|f|^2 + |g|^2 |f - g|^2)} \qquad (3.104)$$

by averaging the intensity difference (numerator) and the total intensity (denominator) separately over the valence energy interval E_2 from -1.0 eV up to the Fermi level. In Fig. 3.71a, we show results for azimuthal angle $\phi = 0°$, for which the plane normal to the reaction plane is a mirror plane. The salient common feature of the experimental and theoretical polarization is the zero crossing at $E_3 - E_4 = 0$. As shown in Fig. 3.70b, c, at this point the entanglement measure S reaches its maximal value 1 and the polarization crosses zero. Thus, the experimental results are therefore consistent with the generation of maximally entangled electron pairs. This is the central result of the work (Vasilyev et al. 2017).

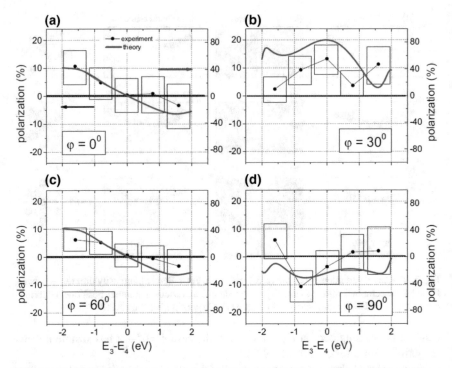

Fig. 3.71 Spin polarization P_3 as a function of the energy difference $(E_3 - E_4)$ for different azimuthal angles ϕ for primary energy $E_1 = 24$ eV. (Reprinted figure with permission from Vasilyev et al. (2017), Copyright (2017) by APS)

As a cross-check, Fig. 3.71c shows experimental data obtained at $\phi = 60°$, which according to symmetry arguments as well as numerical calculations should be identical to those for $\phi = 0°$. The results for $\phi = 30°$ (Fig. 3.71b) and $\phi = 90°$ (Fig. 3.71d), for which the plane normal to the reaction plane is not a mirror plane, are strikingly different. The polarization has no zero crossing, but is positive (negative) throughout for $\phi = 30°$ ($\phi = 90°$).

The above presented theoretical description of the entanglement of electron pairs generated in the (e,2e) reaction at a solid surface was used for the analysis of experimental results of the (e,2e) scattering of spin-polarized electrons from the Shockley surface state of Cu(111). Experimental spin polarization spectra measured for few selected azimuthal orientations of the sample with spin analysis in one of the arms of the (e,2e) spectrometer broadly agree with their theoretical counterparts. They are consistent with spin entanglement of the electron pair at a macroscopic distance from the scattering region. For a definite verification of the entanglement, however, one would need an (e,2e) spectrometer with a spin detector in each arm allowing the measurement of quantum spin correlations and thence the violation of Bell-type inequalities.

References

S.F. Alvarado, R. Feder, H. Hopster, F. Ciccacci, H. Pleyer, Z. Phys. B **49**, 129 (1982)

O.M. Artamonov, S.N. Samarin, J. Kirschner, Appl. Phys. A Mater. Sci. Process. **A65**, 535 (1997)

O.M. Artamonov, O.M. Smirnov, A.N. Terekhov, Phys. Chem. Mech. Surf. **2**(8), 2281–2298 (1985)

O.M. Artamonov, S.N. Samarin, A.N. Vetlugin, I.V. Sokolov, J.F. Williams, J. of Electr. Spectr. and Rel. Phenom. **205**, 66–73 (2015)

I.-G. Baek, H.G. Lee, H.-J. Kim, E. Vescovo, Phys. Rev. B **67**, 075401 (2003)

D.H. Baek, J.W. Chung, W.K. Han, Phys. Rev. B **47**, 8461 (1993)

J. Bansmann, L. Lu, M. Getzlaff, M. Fluchtmann, J. Braun, Surf. Sci. **686**, 454 (2000)

J. Bansmann, M. Getzlaff, G. Schonhense, J. Magn. Magn. Mater. **48**, 60 (1995)

J.S. Bell, Physics **1**, 195 (1964)

C.H. Bennett, D.P. Di Vincenzo, J.A. Smolin, W.K. Wootters, Phys. Rev. A **54**, 3824 (1996)

J. Berakdar, Nucl. Instr. and Meth. in Phys. Res. B **193**, 609 (2002)

J. Berakdar, *Concepts of Highly Excited Electronic Systems* (Wiley-VCH, Weinheim, 2003), p. 189

J. Berakdar, *Electronic Correlation Mapping* (Wiley-VCH, Weinheim, 2006)

J. Berakdar, Phys. Rev. Lett. **83**, 5150 (1999)

J. Berakdar, Phys. Rev. B **58**, 9808 (1998)

J. Berakdar, H. Gollisch, R. Feder, Solid State Commun. **112**, 587 (1999)

H.A. Bethe, E. Salpeter, *Quantum Mechanics of One and Two-Electron Atoms* (Plenum, New York, 1977)

P. Blaha, K. Schwarz, P. Sorantin, S.B. Trickey, Comput. Phys. Commun. **59**, 399 (1990)

K. Blum, *Density Matrix Theory and Applications* (Plenum, New York, 1996)

D. Bohm, Phys. Rev. **85**, 166 (1952)

D. Bouwmeester, A. Ekkert, A. Zeilinger, *The Physics of Quantum Information* (Springer, Berlin, 2000)

P. Bruno, *Physical Origins and Theoretical Models of Magnetic Anisotropy, in Magnetismus von Festkörpern und Grenzflächen*, vol. 24 (Forschungszentrum Jülich, 1993) pp. 1–28

P. Bruno, Phys. Rev. B **39**, 865 (1989)

J.C. Buchholz, M.G. Lagally, J. Vac. Sci. and Technol. **11**(1), 194 (1974)

J.C. Buchholz, M.G. Lagally, Phys. Rev. Lett. **35**, 442–445 (1975)

F. Buscemi, P. Bordone, A. Bertoni, Phys. Rev. A **73**, 052312 (2006)

F. Buscemi, P. Bordone, A. Bertoni, Phys. Rev. A **75**, 032301 (2007)

D.M. Bylander, Leonard Kleinman, Phys. Rev. B **29**, 1534 (1984)

N. Egede Christensen, B. Feuerbacher, Phys. Rev. B **10**, 2349 (1974)

J.F. Clauser, A. Shimony, Rep. Prog. Phys. **41**, 1881 (1978)

R. Cortenraada, S.N. Ermolov, V.N. Semenov, A.W. Denier van der Gon, V.G. Glebovsky, S.I. Bozhko, H.H. Brongersma, J. Cryst. Growth **222**, 154 (2001)

R. Courths, S. Loebus, S. Halilov, T. Scheunemann, H. Gollisch, R. Feder, Phys. Rev. B **60**, 8055 (1999)

R. Courths, M. Lau, T. Scheunemann, H. Gollisch, R. Feder, Phys. Rev. B **63**, 195110 (2001)

M. Donath, J. Phys, Condens. Matter **11**, 9421 (1999)

P. Drude, Ann. Phys. Lpz. **306**, 566–613 (1900)

J.H. Dunn, O. Karis, C. Andersson, D. Arvanitis, R. Carr, I.A. Abrikosov, B. Sanyal, L. Bergqvist, O. Eriksson, Phys. Rev. Lett. **94**, 217202 (2005)

A. Einstein, B. Podolsky, N. Rosen, Phys. Rev. **47**, 777 (1935)

H.J. Elmers, U. Gradmann, Appl. Phys. A **51**, 255 (1990)

M. Farle, A. Berghaus, Yi Li, K. Baberschke, Phys. Rev. B **42**, 4873 (1990)

R. Feder (ed.), *Polarized Electrons in Surface Physics* (World Scientific, Singapore, 1985)

Roland Feder, J. Phys. C: Solid State Phys. **14**, 2049–2091 (1981)

R. Feder, H. Gollisch, Low-energy (e,2e) spectroscopy, in *The Book Solid-State Photoemission and Related Methods*, ed. by W. Schattke, M.A. Van Hove, WILEY-VCH Verlag GmbH&Co.KGaA (2003)

R. Feder, H. Gollisch, D. Meinert, T. Scheunemann, O.M. Artamonov, S.N. Samarin, J. Kirschner, Phys. Rev. B **58**, 16418 (1998)

R. Feder, F. Giebels, H. Gollisch, Phys. Rev. B **92**, 075420 (2015)

R. Feder, H. Gollisch, Solid State Commun. **119**, 625 (2001)

J. Feydt, A. Elbe, H. Engelhard, G. Meister, A. Goldmann, Surf. Sci. **440**, 213–220 (1999)

N. Fominykh, J. Berakdar, J. Henk, P. Bruno, Phys. Rev. Lett. **89**, 086402 (2002)

H. Fritzsche, J. Kohlhepp, U. Gradmann, Phys. Rev. B **51**, 15933 (1995)

O. Fruchart, J.P. Nozières, D. Givord, J. Mag. Magn. Mater. **207**, 158 (1999)

J.C. Fuggle, D. Menzel, Surf. Sci. **53**, 21–34 (1975)

P. Fulde, *Electron Correlations in Molecules and Solids*, Springer Series in Solid State Sciences, vol. 100 (Springer, Berlin, 1993)

G. Garreau, M. Farle, E. Beaurepaire, K. Baberschke, Phys. Rev. B **55**, 330 (1997)

M. Getzlaff, Appl. Phys. A **72**, 455 (2001)

G.C. Ghirardi, L. Marinatto, Phys. Rev. A **70**, 012109 (2004)

F. Giebels, H. Gollisch, R. Feder, Phys. Rev. B **87**, 035124 (2013)

H. Gollisch, Nv Schwartzenberg, R. Feder, Phys. Rev. B **74**, 075407 (2006)

H. Gollisch, R. Feder, J. Phys.: Condens. Matter **16**, 2207 (2004)

H. Gollisch, T. Scheunemann, R. Feder, Solid State Commun. **117**, 691 (2001)

H. Gollisch, Xiao Yi, T. Scheunemann, R. Feder, J. Phys.: Condens. Matter **11**, 9555 (1999)

U. Gradmann, G. Waller, Surf. Sci. **116**, 539 (1982)

N.L. Harshman, G. Hutton, Phys. Rev. A **77**, 042310 (2008)

N.L. Harshman, P. Singh, J. Phys. A: Math. Theor. **41**, 155304 (2008)

M. Hattass, T. Jahnke, S. Schössler, A. Czasch, M. Schöffler, L.P.H. Schmidt, B. Ulrich, O. Jagutzki, F.O. Schumann, C. Winkler, J. Kirschner, J. Dörner, H. Schmidt-Böcking, Phys. Rev. B **77**, 165432 (2008)

F.U. Hillebrecht, A. Morozov, J. Kirschner, Phys. Rev. B **71**, 125406 (2005)

P. Hohenberg, W. Kohn, Phys. Rev. **136**, B864 (1964)

H. Hopster, D.L. Abraham, D.P. Pappas, Spin-polarized EELS of surfaces. J. Electron Spectrosc. Relat. Phenom. **51**, 301 (1990)

R. Horodecki, P. Horodecki, M. Horodecki, K. Horodecki, Rev. Mod. Phys. **81**, 865 (2009)

B.B. Kadomtsev, Phys. Usp. **37**, 425 (1994)

K.P. Kämper, W. Schmitt, D.A. Wesner, G. Güntherodt, Appl. Phys. A **49**, 573 (1989)

Y.K. Kato, R.C. Myers, A.C. Gossard, D.D. Awschalom, Science **306**, 1910 (2004)

J. Kessler, *Polarized Electrons* (Springer, Berlin, Heidelberg, 1985)

A.S. Kheifets, S. Iacobucci, A. Ruoccoa, R. Camilloni, G. Stefani, Phys. Rev. B **57**, 7380 (1998)

K.W. Kim, Y.H. Hyun, R. Gontarz, Y.V. Kudryavtsev, Y.P. Lee, Phys. Status Solidi A **196**(1), 197–200 (2003)

J. Kirschner, *Polarized Electrons at Surfaces* (Springer, Berlin, Heidelberg, New York, Tokyo, 1985a)

J. Kirschner, Phys. Rev. Lett. **55**, 973 (1985b)

J. Kirschner, F. Giebels, H. Gollisch, R. Feder, Phys. Rev. B **88**, 125419 (2013)

J. Kirschner, S. Suga, Surf. Sci. **178**, 327 (1986)

J. Kirschner, D. Rebenstorff, H. Ibach, Phys. Rev. Lett. **53**, 698 (1984)

J. Kirschner, O.M. Artamonov, S.N. Samarin, Phys. Rev. Lett. **75**, 2424 (1995)

J. Kirschner, R. Feder, Phys. Rev. Lett. **42**, 1008 (1979)

H. Knoppe, E. Bauer, Phys. Rev. B **48**(8), 5621 (1993a)

H. Knoppe, E. Bauer, Phys. Rev. B **48**, 1794 (1993b)

W. Kohn, L.J. Sham, Phys. Rev. **140**, A1133 (1965)

W. Kohn, Rev. Mod. Phys. **71**, 1253 (1998)

J. Kolaczkiewiczl, E. Bauer, Surf. Sci. **144**, 495 (1984)

L. Lamata, J. Leon, Phys. Rev. A **73**, 052322 (2006)

L.D. Landau, E.M. Lifshitz, *Quantum Mechanics (Non-relativistic Theory)* (Pergamon Press, 1991)

S. Lee, J.T. Sadowski, H. Jang, J.H. Park, J.Y. Kim, J. Hu, R. Wu, C.C. Kao, Phys. Rev. B **83**, 144420 (2011)

Yi Li, K. Baberschke, Phys. Rev. Lett. **68**, 1208 (1992)

A. Liscio et al., J. Electron Spectrosc. Relat. Phenom. **137–40**, 505 (2004)

H.A. Lorentz, *The Theory of Electrons* (BG Teubner, Leipzig, 1909)

I.E. McCarthy, E. Weigold, Rep. Prog. Phys. **54**, 789 (1991)

J.E. Moore, Nature **464**, 194 (2010)

A. Morozov, J. Berakdar, S.N. Samarin, F.U. Hillebrecht, J. Kirschner, Phys. Rev. B **65**, 104425 (2002)

N.F. Mott, H.S.W. Massey, *The Theory of Atomic Collisions* (Oxford, 1965)

H.P. Oepen, K. Hünlich, J. Kirshner, Phys. Rev. Lett. **56**, 496 (1986)

R. Opila, R. Gomer, Surf. Sci. **105**, 41 (1981)

C.H.F. Peden, N.D. Shinn, Surf. Sci. **312**, 151 (1994)

D.T. Pierce, F. Meier, Phys. Rev. B **13**, 5484 (1976)

G.A. Prinz, G.T. Rado, J.J. Krebs, J. Appl. Phys. **53**, 2087 (1982)

X. Qian, W. Hübner, Phys. Rev. B **60**, 16192 (1999)

X. Qian, W. Hübner, Phys. Rev. B **64**, 092402 (2001)

A. Rampe, D. Hartmann, G. Güntherodt, Linear magnetic dichroism in angle-resolved photoemission spectroscopy from Co(0001) and Fe(110) valence band, in *Proceedings Spin—orbit—influenced Spectroscopies of Magnetic Solids*, ed. by H. Ebert, G. Schuütz (Springer, Herrsching, Germany, 1995)

P. Ravindran, A. Kjekshus, H. Fjelvåg, P. James, L. Nordström, B. Johansson, O. Eriksson, Phys. Rev. B **63**, 144409 (2001)

K.J. Rawlings, Surf. Sci. **99**, 507–522 (1980)

D.M. Riffe, G.K. Wertheim, Surf. Sci. **399**, 248 (1998)

U. Rücker, H. Gollisch, R. Feder, Phys. Rev. B **72**, 214424 (2005)

M.G. Samant, J. Stöhr, S.S.P. Parkin, G.A. Held, B.D. Hermsmeier, F. Herman, M. van Schilfgaarde, L.-C. Duda, D.C. Mancini, N. Wassdahl, R. Nakajima, Phys. Rev. Lett. **72**, 1112 (1994)

S.N. Samarin, Oleg M. Artamonov, Alexander P. Baraban, M. Kostylev, P. Guagliardo, J.F. Williams, Appl. Phys. Lett. **106**, 042404 (2015a)

S. Samarin, O.M. Artamonov, P. Guagliardo, L. Pravica, A. Baraban, F.O. Schumann, J.F. Williams, J. Electron Spectr. Rel. Phenom. **198**, 26–30 (2015b)

S. Samarin, O.M. Artamonov, P. Guagliardo, K. Sudarshan, M. Kostylev, L. Pravica, A. Baraban, J.F. Williams, Surf. Sci. **617**, 22–28 (2013a)

S. Samarin, J. Williams, O. Artamonov, L. Pravica, K. Sudarshan, P. Guagliardo, F. Giebels, H. Gollisch, R. Feder, Appl. Phys. Lett. **102**, 251607 (2013b)

S. Samarin, O. Artamonov, J. Berakdar, A. Morozov, J. Kirschner, Surf. Sci. **482–485**, 1015–1020 (2001)

S.N. Samarin, O.M. Artamonov, A.D. Sergeant, J.F. Williams, Surf. Sci. **601**, 4343 (2007a)

S.N. Samarin, J.F. Williams, A.D. Sergeant, O.M. Artamonov, H. Gollisch, R. Feder, Spin-dependent reflection of very-low-energy electrons from W(110). Phys. Rev. B **76**, 125402 (2007b)

S. Samarin, O.M. Artamonov, A.D. Sergeant, R. Stamps, J.F. Williams, Phys. Rev. Lett. **97**, 096402 (2006)

S.N. Samarin, J. Berakdar, O. Artamonov, J. Kirschner, Phys. Rev. Lett. **85**, 1746 (2000a)

S. Samarin, J. Berakdar, O. Artamonov, H. Schwabe, J. Kirschner, Surf. Sci. **470**, 141 (2000b)

S. Samarin, O.M. Artamonov, A.D. Sergeant, J.F. Williams, Surf. Sci. **579**, 166 (2005a)

S.N. Samarin, O.M. Artamonov, A.D. Sergeant, J.F. Williams, Phys. Rev. B **72** 235419/1–5 (2005b)

S. Samarin, R. Herrmann, R. Schwabe, O. Artamonov, J. Electron Spectrosc. Relat. Phenom. **96**, 61 (1998)

S. Samarin, J. Berakdar, R. Herrmann, H. Schwabe, O. Artamonov, J. Kirschner, J. Phys. IV France 9 Pr6-137–Pr6-143 (1999)

S. Samarin, O.M. Artamonov, A.D. Sergeant, J. Kirschner, A. Morozov, J.F. Williams, Phys. Rev. B **70**, 073403 (2004)

S. Samarin, O.M. Artamonov, V.N. Petrov, M. Kostylev, L. Pravica, A. Baraban, J.F. Williams, Phys. Rev. B **84**, 184433 (2011a)

S.N. Samarin, O.M. Artamonov, A.D. Sergeant, L. Pravica, D. Cvejanovic, P. Wilkie, P. Guagliardo, A.A. Suvorova, J.F. Williams, J. Phys.: Conf. Ser. **288**, 012015 (2011b)

S. Samarin, O.M. Artamonov, A.D. Sergeant, J. Kirschner, J.F. Williams, J. Phys.: Conf. Ser. **100**, 072033 (2008)

D. Sander, A. Enders, C. Schmidthals, J. Kirschner, H.L. Johnston, C.S. Arnold, D. Venus, J. Appl. Phys. **81**, 4702 (1997)

D. Sander, C. Schmidthals, A. Enders, J. Kirschner, Phys. Rev. B **57**, 1406 (1998)

D. Sander, J. Phys, Condens. Matter **16**, R603–R636 (2004)

D.S. Saraga, B.L. Altshuler, D. Loss, R.M. Westervelt, Phys. Rev. Lett. **92**, 246803 (2004)

B. Schirmer, M. Wuttig, Phys. Rev. B **60**, 12945 (1999)

F.O. Schumann, C. Winkler, J. Kirschner, Phys. Rev. B **88**, 085129 (2013)

F.O. Schumann, C. Winkler, J. Kirschner, Phys. Rev. Lett. **98**, 257604 (2007a)

F.O. Schumann, C. Winkler, J. Kirschner, New J. Phys. **9**, 372 (2007b)

F.O. Schumann, J. Kirschner, J. Berakdar, Phys. Rev. Lett. **95**, 117601 (2005)

F.O. Schumann, C. Winkler, J. Kirschner, F. Giebels, H. Gollisch, R. Feder, Phys. Rev. Lett. **104**, 087602 (2010)

J. Schliemann, J.I. Cirac, M. Kus, M. Lewenstein, D. Loss, Phys. Rev. A **64**, 022303 (2001)

A.M. Shikin, O. Rader, G.V. Prudnikova, V.K. Adamchuk, W. Gudat, Phys. Rev. B **65**, 075403 (2002)

J.C. Slater, Rev. Mod. Phys. **6**, 209 (1934)

A. Sommerfeld, Naturwissenschaften **15**, 63 (1927)

M. Streun, G. Baum, W. Blask, J. Berakdar, Phys. Rev. A **59**, R4109 (1999)

J. Stöhr, J. Magn. Magn. Mater. **200**, 470 (1999)

M.C. Tringides, Phys. Rev. Lett. **65**, 1372–1375 (1990)

C. Tusche, A. Krasyuk, J. Kirschner, Ultramicroscopy **159**, 520 (2015)

D. Vasilyev, F.O. Schumann, F. Giebels, H. Gollisch, J. Kirschner, R. Feder, Phys. Rev. B **95**, 115134 (2017)

D. Vasilyev, C. Tusche, F. Giebels, H. Gollisch, R. Feder, J. Kirschner, J. Electron Spectrosc. Relat. Phenom. **199**, 10 (2015)

D. Venus, S. Cool, M. Plihal, Surf. Sci. **446**, 199 (2000)

J. Wang, C.K. Law, M.C. Chu, Phys. Rev. A **73**, 034302 (2006)

E. Wigner, F. Seitz, Phys. Rev. **43**, 804 (1933)

J.F. Williams, S.N. Samarin, O.M. Artamonov, A.D. Sergeant, L. Pravica, D. Cvejanovic, P. Wilkie, J. Phys.: Conf. Ser. **80**, 012024 (2007)

S.A. Wolf, D.D. Awschalom, R.A. Buhrman, J.M. Daughton, S. von Molnàr, M.L. Roukes, A.Y. Chtchelkanova, D.M. Treger, Science **294**, 1488 (2001)

P.K. Wu, M.C. Tringides, M.G. Lagally, Phys. Rev. B **39**, 7595 (1989)

J. Wunderlich, B. Kaestner, J. Sinova, T. Jungwirth, Experimental observation of the spin-Hall effect in a two dimensional spin–orbit coupled semiconductor system. Phys. Rev. Lett. **94**, 047204 (2005)

Y. Yamamoto, T. Miura, M. Suzuki, N. Kawamura, H. Miyagawa, T. Nakamura, K. Kobayashi, T. Teranishi, H. Hori, Phys. Rev. Lett. **93**, 116801-1 (2004)

R.X. Ynzunza, H. Daimon, F.J. Palomares, E.D. Tober, Z. Wang, F.J.G. de Abajo, J. Morais, R. Denecke, J.B. Kortright, Z. Hussain, M.A. Van Hove, C.S. Fadley, J. El. Spectr. Rel. Phenom. **106**, 7 (2000a)

R.X. Ynzunza, R. Denecke, F.J. Palomares, J. Morais, E.D. Tober, Z. Wang, F.J. Garcia de Abajo, J. Liesegang, Z. Hussain, M.A. Van Hove, C.S. Fadley, Surf. Sci. **459**, 69–92 (2000b)

K. Zakeri, T.R.F. Peixoto, Y. Zhang, J. Prokop, J. Kirschner, Surf. Sci. **604**, L1 (2010)

M.A. Zaluska-Kotur, S. Krukowski, Z. Romanowski, L.A. Turski, (2002) Phys. Rev. B **65** 045404/1-9

H. Zillgen, B. Feldmann, M. Wuttig, Surf. Sci. **321**, 32 (1994)

I. Žutć, J. Fabian, S.D. Sarma, Rev. Mod. Phys. **76**, 323 (2004)

Chapter 4
Emission of Correlated Electron Pairs from Surfaces Induced by Photons, Positrons and Ions

Abstract The phenomena associated with the emission of two time-correlated electrons from a solid surface are expanded using incident photons, positrons, and ions. The influence of particle interactions on electron correlation in surfaces and thin films is indicated.

Emission of correlated electron pairs from solid surfaces upon electron impact was discussed above. However, there are other means to supply energy to the solid to release a pair of electrons from its surface, for example, by a photon, positron or via neutralization of an ion in the vicinity of a solid surface. These processes, underlying mechanisms and possible information that can be gained by analyzing energy distributions of two emitted electrons will be briefly discussed in this section. Perhaps future studies may extend the quantum descriptions to incident uncharged and charged atomic or molecular ions on surfaces as in hydrogen fusion devices.

4.1 Emission of Correlated Electron Pairs from a Solid Surface upon Single Photon Absorption

As was above mentioned two-electron emission from a solid surface induced by a low-energy incident electron carries information on the electron-electron correlation. Another spectroscopic method for electronic correlation is the coincident, photon-induced electron-pair emission (γ,2e) or Double Photo-Emission (DPE), i.e. emission of correlated electron pairs following single-photon absorption by a surface. This technique is even more sensitive to the effects of correlations than (e,2e) spectroscopy. In fact, it can be shown that the cross section for (γ,2e) vanishes in the absence of correlations (Berakdar 1998). A competing mechanism for generation of pairs of electrons may proceed in two steps, namely a conventional photoemission (or photoionization) event leading to a single energetic free electron, followed by an inelastic collision where part of the energy of this electron is transferred to a second electron site or different atom/lattice sites. That process also leads to the simulta-

© Springer Nature Switzerland AG 2018

S. Samarin et al., *Spin-Polarized Two-Electron Spectroscopy of Surfaces*, Springer Series in Surface Sciences 67, https://doi.org/10.1007/978-3-030-00657-0_4

neous emission of two electrons from the target, and there is no way to eliminate experimentally a background from such processes from the measured signal, e.g., by time resolution. In that way two-electron photoemission on solids is different from studies on gas phase targets, where the relatively low density of the target atoms or molecules usually ensures that if two electrons are detected in coincidence they originate from the same target particle, and therefore should display the broadest influence of electronic correlation in a double ionization event.

Experimentally, the realization of the $(\gamma, 2e)$ process at surfaces has long been a challenge due to the low count rates (compared to single electron emission) which is inherent to coincidence studies. First experiments with a momentum resolution of the photoelectrons were reported many years ago (Herrmann et al. 1998).

The concept of an electron pair "quasiparticle" was suggested for the interpretation of DPE, which can be regarded as single photoemission of a "quasiparticle" formed by an electron pair (Berakdar 1998) and was also first introduced to interpret diffraction patterns observed in the (e,2e) spectrum of a crystalline surface (Berakdar 1998). In addition to the known features of Angle-resolved Ultraviolet Photo-Electron Spectroscopy (ARUPS), the spectra of this photoemitted "quasiparticle" reveal a dependence on the pair's internal degree of freedom that characterizes the mutual interaction of the two emitted electrons. Consequently DPE experiments have been expected to provide a direct insight into the influence of electronic correlation on initial and final many-body states.

The theoretical description of double-electron photoemission from solid surfaces identified two cases: (i) a pair of electrons are excited from localized electronic states and (ii) electrons are excited by an incident photon from delocalized electronic states (Berakdar 1998; Fominikh et al. 2001). In both cases the matrix element M_{fi} describing transition from initial (bound) state to the final state of two outgoing electrons with wave vectors $\mathbf{k}_a$ and $\mathbf{k}_b$ is proportional to the scalar product: $M_{fi} \propto (\boldsymbol{\varepsilon} \cdot \mathbf{k}^+)$, where $\boldsymbol{\varepsilon}$ is polarization vector of the incident photon and $\mathbf{k}^+ = \mathbf{k}_a + \mathbf{k}_b$ the total momentum of the pair. The relation $M_{fi} \propto (\boldsymbol{\varepsilon} \cdot \mathbf{k}^+)$ can be considered as propensity rule. The condition $(\boldsymbol{\varepsilon} \cdot \mathbf{k}^+) = 0$ (selection rule) means that the emission of the two-electron pair excited by a photon is not allowed when the polarization of the incident photon beam is perpendicular to the momentum of the centre of mass of the pair $\mathbf{k}^+$.

The first experiment on a double photoemission from metal surfaces (Herrmann et al. 1998) was performed at the electron storage ring BESSY in Berlin. The energy and momenta of two electrons emitted simultaneously from the (001) surfaces of a Cu and Ni samples after absorption of 45 eV photons were measured. Schematic of the experimental set-up is shown in Fig. 4.1. Photons with energy of 45 eV impinge on the sample surface at normal incidence. The polarisation vector $\boldsymbol{\varepsilon}$ is in the scattering plane containing the normal to the surface and two detectors D1 and D2, located at $\pm 40°$ with respect to the normal. The energy dispersed, p-polarized synchrotron radiation from a toroidal grating monochromator passed through an aperture of 30 μm diameter onto the sample surface. The photon beam was time-structured, in a "single bunch" mode; i.e., the intensity was concentrated in regular bunches of 0.6 ns half-width and 200 ns time distance. The mean beam intensity was adjusted so that

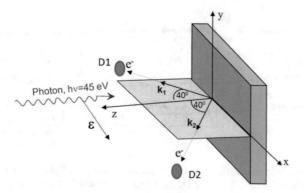

Fig. 4.1 Schematic of the double electron photoemission experiment

the average number of photons per bunch was less than one to ensure unique detection per incident photon. While the Cu sample was a pure Cu crystal, the Ni sample was produced by evaporating approximately 10 monolayers of Ni on top of the Cu(001) surface. Because of the small escape depth of electron pairs in the energy range considered (<5 monolayers), it is assumed the electron emission characteristics from the epitaxial Ni layer to be equivalent to that from a bulk (100) Ni crystal.

Figure 4.2 represents results of the double photoemission measurements on the Cu(001) sample. A density plot of the two dimensional energy distribution for correlated electron pairs originating from a Cu(001) surface following the absorption of 45 eV photons is shown in Fig. 4.2a, the total energy distribution of electron pairs is presented in Fig. 4.2b, and the energy sharing distribution of electron pair with total energy within 2 eV band below the Fermi level is shown in Fig. 4.2c. From Fig. 4.2a, c it is evident that the maximum sum energy of correlated electron pairs is given by $E_{sum} = 35$ eV (shown by dashed line in Fig. 4.2a. Considering the photon energy of 45 eV and the work function in Cu of approximately 5 eV, a constant sum energy of $E_{sum} = 35$ eV represents electron pairs emitted from the vicinity of the Fermi level (Fig. 4.3). Indeed, for the two electrons excited from the vicinity of the Fermi level the following energy balance is valid: $(E_1 + E_2)_{max} = h\nu - 2e\varphi$, where $h\nu$ is the photon energy and $e\varphi$ is the work function of the surface.

It should be noticed here that coincidences between photoelectrons from core levels and subsequent Auger electrons cannot explain the intensity distribution in the spectrum of Fig. 4.2a. Indeed, photoelectrons from core levels have discrete energies and Auger electrons from the valence band may be distributed in energy over a maximum range of about 6 eV according to the valence band width of Cu (and Ni). Considering the experimental energy resolution, coincidences between photo- and Auger electrons would show up in the two-dimensional energy spectrum as isolated elliptical intensity spots, having a width of about 1–2 eV in the one direction and about 6 eV in the other direction. Then the data of Fig. 4.2 can be interpreted as clear experimental evidence for the simultaneous excitation of two electrons from the conduction band of Cu by the absorption of one photon.

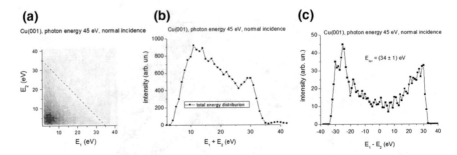

(a) Cu(001), photon energy 45 eV, normal incidence

(b) Cu(001), photon energy 45 eV, normal incidence

(c) Cu(001), photon energy 45 eV, normal incidence

$E_{tot} = (34 \pm 1)$ eV

Fig. 4.2 Double photoemission from Cu(001) surface excited by 45 eV photons: **a** two-dimensional energy distribution of correlated electron pairs; **b** total energy distribution; **c** energy sharing distribution of pair with total energy $E_{tot} = (34 \pm 1)$ eV

Fig. 4.3 Energy balance in double electron photoemission

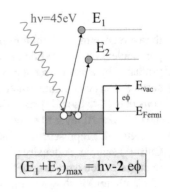

$$(E_1 + E_2)_{max} = h\nu - 2\ e\phi$$

In addition to coincidence events, a background contribution of electron pairs was measured including single photoelectrons from the valence band that did not undergo considerable energy loss in random coincidence with various electron energies in the second detector. Furthermore, one can see in Fig. 4.2a a strong contribution of electron pairs with summed energies of only a few eV. It is assumed that those electrons originate from various energy loss processes of single photoelectrons and correlated pairs in the solid. Further analysis is concentrated on electron pairs with summed energies close to $(E_{sum})_{max} = 35$ eV, since their kinematics results from the nearly pure two-electron photoexcitation from the vicinity of the Fermi level with almost no perturbations from energy loss processes. Figure 4.2b, c show the total energy distribution and the energy sharing distribution of correlated electron pairs, respectively. Cut-off at 35 eV on the total energy distribution indicates, as was mentioned, the maximum total energy of pairs that corresponds to the emission of electron pairs from the vicinity of the Fermi level. The energy sharing distribution covers an energy sharing range $(E_1 - E_2)$ between −33 eV and +33 eV, which does not include correlated pairs in which one electron carries less than 3 eV. This restriction is a consequence of the experimental lower detection limit of 2−3 eV and does not reflect the physical probability for the appearance of such unequal sharing. Indeed, because of the photon bunch repetition rate of 1/200 ns⁻¹, the maximum flight time (relative to the elastic electron flight time) that could be measured was 200 ns, corresponding to a minimum kinetic energy of 2−3 eV. In the energy sharing distribution a minimum for equal energy sharing ($E_1 - E_2 = 0$) appears, indicating a higher probability for the two electrons of a pair to be emitted with unequal energies into the solid angles viewed by the detectors, rather than sharing their sum energy equally. This experimental finding is consistent with the propensity rule (Berakdar 1998). It was shown that the dipole transition amplitude for the single step photo double ionization vanishes when the light polarization vector is perpendicular to the sum momentum vector of the two emitted electrons. This is the case for symmetric emission angles and equal energies of both outgoing electrons since the polarization vector of the light is parallel to the sample surface (at normal incidence of the light). In that case the theoretical cross section of the single step, photo double ionization process is zero, whereas the corresponding theoretical cross section of the two step mechanism remains finite (Berakdar 1998). Given the detection geometry is symmetric the minimum at equal energies in the energy sharing distribution of Fig. 4.2c could be interpreted as a signature of the contribution of the single step photo double ionization mechanism. On the other hand it is noted that the high angular acceptance of electron detectors always measured a superposition of symmetric and asymmetric emission angles. Therefore, the contribution of measured pairs with equal electron energies (integrated over the acceptance angles) is always finite even if the cross section for symmetric emission of those pairs vanishes. For the Ni(001) sample, the same main features in the two-dimensional time-of-flight and energy distributions are observed and a similar analysis as described for Cu leads to the equivalent conclusion of a two-electron photoemission process in the conduction band of Ni. However, the depth of the minimum in the energy sharing distribution of electron pairs excited from Cu(001) is significantly larger than in Ni(001) case. Following the above interpretation, one can take this finding as an indication for a higher contribution of the two-step photo double ionization process as underlying mechanism for the emission of electron pairs from the vicinity of the Fermi level of Ni than in the corresponding case of Cu.

Correlation between electrons may show up both in the angular and in the energies distribution of the two emitted electrons following the absorption of a photon. The first type of correlation was demonstrated (Schumann et al. 2007), where the authors investigated the electron pair emission from a Cu(111) surface upon photon absorption. They were able to observe, for the first time, the full extension and shape of a depletion zone around the fixed emission direction of one electron with an angular extension of −1.2 rad, which was independent of the electron energy. For NaCl excited by 34 eV photons, the diameter of the reduced intensity region in the second electron momentum distribution was found to be −1.1 Å^{-1} (Schumann et al. 2006) for a free electron energy of 9.5 eV and the second electron with fixed 5.5 eV energy in the detection direction. In that case the diameter was significantly smaller than the theoretical value for Cu (Fominykh et al. 2002). A qualitatively similar momentum distribution was measured for the excitation of LiF by electron impact (Schumann et al. 2006).

From the energy point of view, the correlation is characterized by a correlation energy U defined as $U = E_{N-2} - E_{N-1,1} - E_{N-1,2}$ (Hillebrecht et al. 2005) where E_{N-2} is the energy of the system with two holes on one atom, i.e., the final state generated in our experiment. $E_{N-1,i}$ are the energies of the system with only one hole on site i. The correlation energy describes the energy difference between a final state with two holes, generated at the same time on the same site, such that the holes interact with each other via Coulomb repulsion, and it is this correlation energy which can be determined from $(\gamma,2e)$ experiment. Single holes are in general not correlated in time or space and two holes may be produced which are too far apart in space and time to interact with each other, and which therefore are uncorrelated. The correlation energy is expected to be present in all processes leading to a two holes final state, i.e., emission of a pair of electrons from one atom or lattice site caused by the absorption of a single photon. These processes are called one step events.

The first experimental observation of the correlation energy U using $(\gamma,2e)$ spectroscopy (Hillebrecht et al. 2005) was carried out on C_{60} films (between one and ten monolayers) on Cu(111) substrate. The $(\gamma,2e)$ spectra were recorded at the beamline G1 of Hasylab, Hamburg using Time-of-Flight electron energy analysis. The time resolution of the experiment was of the order of 0.3 ns, yielding an energy resolution of about 0.5 eV for the lowest energies. The two-electron spectra were measured at 45 eV photon energy with polarization of the light in the scattering plane at two angles of incidence: 0° and 35° with respect to the surface normal. Energy sharing distributions measured at different angles of incidence are almost identical.

For the discussion of the $(\gamma,2e)$ results on C_{60} films, a binding energy is defined for two electrons in a fashion analogous to conventional photoemission, instead of considering the total kinetic energy. The two-electron binding energy (2eBE) is given by

$$2eBE = h\nu - E_{k,1} - E_{k,2} - 2e\varphi,$$

where $E_{k,1}$ and $E_{k,2}$ are the kinetic energies of electrons, $e\varphi$ is the work function of the surface, and $h\nu$ is the photon energy. The onset of the distribution of two-electron binding energies corresponds to a final state where both electrons have been ejected from the Fermi level E_F, i.e., from the highest occupied level. Since each electron must overcome the sample work function $e\varphi$, the onset occurs at $(E_{k,1} + E_{k,2}) = h\nu - 2e\varphi$, which corresponds to 2eBE = 0.

In a simple picture, the probability to find a certain two-electron binding energy is expected to be related to the density of occupied electronic states. In single photoemission, the spectral intensity for a given kinetic or binding energy is given by the joint (i.e., initial and final) density of states, multiplied by the appropriate matrix element. In double photoemission, one can start from an analogy to a single photoemission. Then the spectral intensity for particular sum energy should be given by all combinations of single photoemission events which yield that particular 2eBE. It is clear that this is represented by the self-convolution of the single photoemission spectrum. The single photoemission spectrum of C_{60} (Tsuei et al. 1997; Hoogenboom et al. 1998) shows several features which are well understood in terms of electronic structure calculations and associated with certain molecular orbitals. In contrast to the single photoemission data, the experimental $(\gamma,2e)$ spectra at first sight do not show structures which appear to be related to features in the density of states. However, careful inspection (after removing background) suggests that at least the spectrum for the monolayer may contain some weak features (Fig. 4 in Hillebrecht et al. 2005). The spectrum for the monolayer of C_{60} shows some fine structure superimposed on a rising spectrum. In contrast, the spectrum for the thick film does not show systematic modulations outside the statistical uncertainty.

It was suggested (Hillebrecht et al. 2005) that the final states for double photoemission and the carbon *KVV* Auger transition are identical because in both cases there are two interacting holes in the valence band. Therefore, one can expect to find similar features in the spectra. Indeed, it turns out that there is a close correspondence between the fine structure observed in the $(\gamma,2e)$ and the *KVV* Auger spectra of a C_{60} monolayer on Cu(111). Closer inspection shows that while the positions of the peaks in the $(\gamma,2e)$ spectrum correspond to those of the *KVV* Auger spectrum, the intensities are not the same. A possible explanation is that the transition matrix elements for the $(\gamma,2e)$ two-hole final states are not identical to those in the Auger process.

A comparison of a self-convolution of the density of states as measured by single photoemission with the $(\gamma,2e)$ or Auger spectrum shows the same sequence of peaks as the $(\gamma,2e)$ or Auger spectrum. However, in order to make the peak positions coincide one must shift the self-convolution spectrum by 1.6 eV towards higher 2eBE (similar to Lof et al. 1992).This shift is attributed to electron correlation on the basis of more energy required to generate two vacancies on one lattice site, in this case within one molecule rather than to eject two electrons independently from each other. The observation of this correlation shift demonstrates that at least a fraction of the observed double photoemission is associated with events in which two electrons are ejected from one lattice site, or in the case of C_{60} from within a single atom within one molecule. This is the first direct observation of the correlation energy in double photoemission from a solid target.

Experimental observation of the Double Electron Photoemission using new time-of-flight imaging technique for pairs of electrons emitted in coincidence from a surface ("Reaction microscope") (Hattass et al. 2008) confirmed manifestation of the "exchange—correlation hole" in the angular distribution and in momentum distribution of pairs. The propensity rule for the DPE from surfaces was checked out evaluating experimental data in terms of "quasiparticles" characterized by K^+ and K^- (Berakdar 1998-I).

Indeed, according to this rule the intensity of DPE is proportional to the product $(K^+ \cdot \varepsilon)$, where ε is the polarization vector of the incident light. The reaction microscope with combined electric and magnetic fields allows for the detection of electrons in the energy range of interest over the full 2π hemisphere above the sample (Hattass et al. 2004). All information for each event is stored in an event-mode data acquisition system. In that way it is possible to construct every projection and cut through the data in an off-line data analysis. Using these capabilities, the data were filtered in two ways. First, the sum momentum $K^+ = k_1 + k_2$ is oriented parallel to

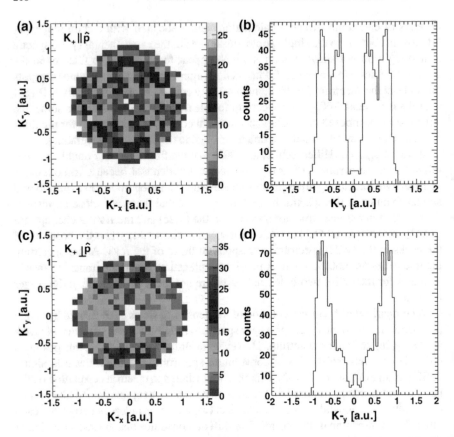

Fig. 4.4 Distribution of relative momenta for different orientations of the sum momentum K^+ with respect to the polarization vector ε. $h\nu$ is 40 eV and sum energies $E_1+E_2 = h\nu -2\,W_\varphi \pm 3$ eV, equal energy sharing. **a** Sum momentum parallel to polarization and **c** sum momentum perpendicular to polarization. **b** and **d** are projections of the area $K_x^- \pm 0.3$ a.u. to the Y axis. (Reprinted figure with permission from Hattass et al. (2008), Copyright (2008) by APS)

the polarization vector of the light and should contain most events that have been created by the DPE process.

Secondly, the sum momentum is perpendicular to the polarization vector, which should lead to a vanishing DPE intensity according to the propensity rule. Comparing the distributions of correlated electron pairs corresponding to these two cases (Fig. 4.4) the conclusion was made: undoubtedly, there is a dependency on the light polarization in the pairs distribution, but a simple explanation by means of the propensity rule is not sufficient.

Another $(\gamma,2e)$ experiment (Trützschler et al. 2017) allowed identification of band-resolved correlated electron pairs, i.e. pairs consisting of two d electrons (d-d pairs) as well as pairs with one sp and one d electron (sp-d pairs). They were resolved in the two-dimensional energy spectrum of Ag(001) and Cu(111) excited by a high-order harmonic of a laser light source (Huth et al. 2014; Chiang et al. 2015). In their measurement two time-of-flight energy analysers in combination with a pulsed light source allowed observation of the band-dependent signatures of the two-particle spectra of Ag and Cu. Pairs of interacting valence electrons were identified according to their summed kinetic energies (E_{sum}) and specifically related to the number of participating d electrons. These two-electron step-wise features in the E_{sum} distribution constitute a more intricate structure (see Fig. 4.5) than the self-convoluted single particle density of states and provide evidence for a distinctly band-dependent electron correlation in these metals. Because E_{sum} is an appropriate quantum number for an electron pair instead of their individual energies, the DPE results provide more information regarding the electron pair configurations which go beyond the capabilities of single-particle spectroscopies. Since the position of characteristic steps in the (γ,2e) binding energy spectrum of Ag(001) corresponding to *sp-sp, sp-d, d-d* pairs excitation does not depend on the photon energy (Fig. 4.5) one can associate these features with features of occupied two-electron states that can be specified only by their total energy as a proper quantum number of a two particle system.

In spite of similar single-particle band structures of Ag and Cu, they show remarkably different response in the two-particle spectrum. The emission of *sp-d* electron pairs is enhanced in Ag whereas barely observable in Cu due to the weaker *sp-d* interaction. Moreover, the energy sharing within a pair depends sensitively on the constituent valence electrons and provides hints of possible energy exchange between electrons via a plasmon in Ag or via the on-site Coulomb interaction in Cu (Trützschler et al. 2017).

4.2 Positron-Induced Positron-Electron Pairs Emission from Surfaces

In a pursuit of studying electron correlation in solids it is very natural to arrive to the idea of using positrons as incident projectiles while detecting in coincidence the scattered positron and an electron resulting from the incident positron scattering on a valence electron of a solid surfaces.

When a positron impinges on a surface one can observe the following scenarios (Fig. 4.6): 1—incident positron can undergo an elastic or inelastic scattering from a surface, escape the surface and be detected by a detector; 2—incident positron can excite a valence electron of the surface and this electron escapes the surface and is detected by a detector; 3—incident positron can be scattered by a valence electron of the surface and then both electron and the positron escape the surface and are detected by two detectors coincidently; 4—incident positron can excite two electrons from the surface and these electrons are detected by two detectors in coincidence; 5—process similar to (3), but now positron and electron detectors are interchanged. Note that the process when two positrons are detected in coincidence (with coincidence window about 200 ns) is impossible. Indeed, given the realistic magnitude of the incident

Fig. 4.5 An overview of the band-specific electron pairs excitation from solid surfaces by photons. **a** DPE data at $h\nu = 32.3$ eV on Ag(001). **b** E_{sum} spectrum from (**a**) integrated over $E_{diff} = \pm 1$ eV. **c** The same as (**b**) for $h\nu = 25.1$ eV. DOS of Ag with DPE processes. E_V is the vacuum level. (Reprinted figure with permission from Trützschler et al. (2017), Copyright (2017) by APS)

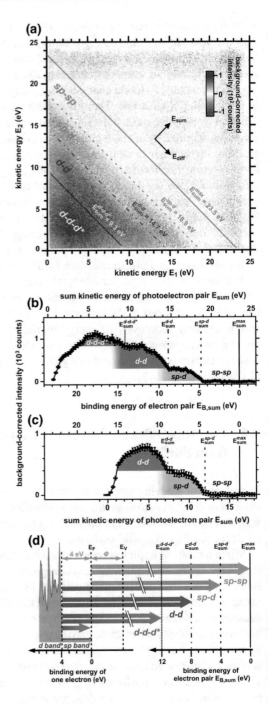

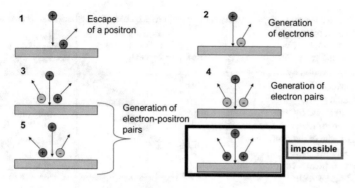

Fig. 4.6 Scenarios of positron-surface scattering with emission of one or two particles

positron current is on the order of 1000 s^{-1}, means that the average time distance between positrons is about 1 ms. So, even accidental coincidence detection of two independently scatted positrons in the coincidence window of 200 ns is very unlikely. We focus now on the reaction depicted in Fig. 4.6 (3) and (5).

If a primary positron hits a surface it can lead to the emission of a positron-electron pair that is termed as (p,ep) reaction. This process was analysed theoretically (Giebels et al. 2009; Berakdar 2000) and observed experimentally (van Riessen et al. 2008).

Correlation of electrons in solids includes two components: Coulomb interaction and electron exchange (due to indistinguishability of electrons). Positron and electron are distinguishable, hence the interaction between them does not include exchange. This feature allows demonstration of pure Coulomb correlation without the exchange component. So, disentangling these effects is possible when one of the interacting electrons is replaced with its antiparticle, the positron. This can be realized experimentally by measuring the momenta of an electron and positron emitted from a surface upon positron impact (van Riessen et al. 2009; Brandt et al. 2015). It was demonstrated that the positron-electron pair production at surfaces leads to features in the energy distributions which are a consequence of distinguishable collision partners (Brand et al. 2014). Specifically, the energy sharing curves are asymmetric with respect to the zero point, where the energy of an electron and a positron of the pair are equal (Fig. 4.7). Symmetric emission geometry was employed via a normal incident positron beam and symmetric detection of an electron and a positron. Using fundamental symmetry arguments the asymmetry of sharing distributions was qualitatively predicted (Brandt et al. 2014). These arguments cannot be used if the emission geometry is broken by a non-normal incident beam. The comparison of data measured at off-normal incidence with the data obtained at normal incidence shows that the change in the symmetry hardly affects the (p,ep) energy sharing distributions. For a variety of materials (p,ep) spectra show a consistent behaviour in the sense that the positron carries on average the larger part of the available energy.

Fig. 4.7 Energy sharing curves obtained by fixing the sum energy to 30 ± 2 eV. The Ag, Co, and NiO samples were excited by 42 eV positrons. The insets indicate which of the particles is detected by the respective spectrometer. Additionally, the curve from the corresponding (e,2e) experiments is presented. (Reprinted figure with permission from Brandt et al. (2014), Copyright (2014) by APS)

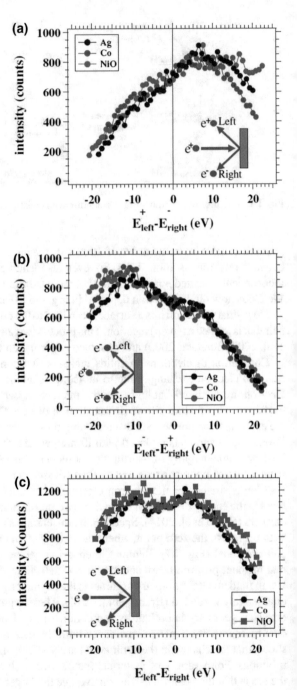

Comparing (p,ep) reaction in several metals (Ag, Co, Cu, Ni) and NiO it was found that the pair emission rate for NiO is enhanced by a factor 2–3 compared to the metals. It is proposed that the positron-electron coupling increases in strongly correlated materials (Brandt et al. 2015).

4.3 Ion-Induced Correlated Electron Pairs Emission from Surfaces

Correlation of electrons in atoms, molecules and solids is characterized by a correlation energy and by a correlation time. The measure of correlation energy is the difference between the energy which is required to remove two electrons from the system (solid) "simultaneously" and summed energy for removal of two electrons successively (one-by-one). As mentioned above (see "Double photoemission from solid surface") the two-electron correlation energy in a solid can be estimated using double photoemission, i.e. correlated emission of two electrons induced by absorption of a single photon. It turns out that a characteristic time scale for the correlated electron dynamics in the metal can be investigated using coincidence spectroscopy with detection of electron pairs that originate from a correlated single step neutralization of the He^{2+} ions (Li et al. 2017).

The correlation time T_{corr} is the time required by the system to respond to external perturbation. When an electron is removed from a metal (as a result of a photoemission, for example), the system will re-arrange in response to such a perturbation. If the next electron is removed by a photon excitation, it will take the same amount of energy as required to remove the first electron. However, if the second photon arrives within time $\tau \ll T_{corr}$ after the first photon, the process occurs according to a different scenario. To illustrate the difference between "sequential" and "simultaneous" double ionization let us consider the He atom. It may display double ionization if the photon energy exceeds the double ionization energy, DI = 79.01 eV. The energy in excess of this threshold is carried away by the two liberated electrons. The energy sum of these two electrons is fixed due to energy conservation, but it is continuously shared among them (Achler et al. 2001). A He^{2+} ion can also be generated by the sequential absorption of two photons if they exceed the first and second ionization energy of 24.59 and 54.42 eV, respectively. In contrast to single-photon ionization, the emitted electrons now carry well-defined energies while possessing an energy sum determined by energy conservation.

The availability of intense and short light pulses allows addressing fundamental questions on the time evolution of the electron dynamics leading to electron emission. For example, if two photons "close in time" excite a sample, will the system react as if two independent energy quanta are absorbed, or does it recognize the two photons as one energy quantum? Theory answered this question analyzing experimental results of two-photon double ionization of He via an intense photon beam with 70 eV photons (Feist et al. 2009a, b). A single 70 eV photon cannot create a He^{2+} ion because of

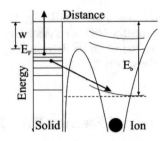

Fig. 4.8 Neutralization step via Auger capture. E_F (W) denotes the Fermi level (work function) of the metal. The energy level of the projectile at infinite distance is E_b. The atomic levels are shifted close to the surface due to the image charge. A metal electron makes a transition into the lower lying level of the ion leading to electron emission from the metal. (Reprinted figure with permission from Li et al. (2017), Copyright (2017) by APS)

the energy balance, only the absorption of at least two photons can achieve this. If the temporal width of the beam is below 300 attosec, most of the two emitted electrons share the available energy continuously. Hence, the system responds as if one energy quantum of 140 eV is absorbed. Pulse widths larger than 4.5 fs lead to electron emission with well-defined energies indicative of two sequential ionization steps. Such studies hold the promise to determine the time scale τ for correlated electron dynamics.

It was demonstrated (Li et al. 2017) that an estimate of the electron correlation time at a metal surface can be obtained using an alternative route based on the reversed process of ionization called neutralization near a surface. This effect takes place about 2–6 Å from the surface and causes electron emission via a non-radiative transfer of energy. It is an efficient process, because almost all incoming He^{2+} ions leave the surface region as atoms (Oliphant 1930). Transition rates for He^{2+} neutralization are known (Lorente and Monreal 1994; Wethekam et al. 2003) and correspond to an average time of 2–20 femtosec between the two neutralization steps.

There is a finite probability for two energy quanta to become available covering the interesting range of 300 as to 4.5 fs (Li et al. 2017). Figure 4.8 shows one particular neutralization path for a singly charged ion leading to electron emission. As the ion approaches the metal surface the atomic energy levels are bent slightly upwards due to the image charge of the ion and with a shift of about 2 eV (Hagstrum 1966). Eventually an electron in the metal gains enough energy to be emitted in an Auger capture process.

For a doubly charged He ion a sequence of different paths leads to a neutral atom (Lorente and Monreal 1994). For low kinetic energy He ions the kinetic energy can be ignored in the energy balance (Hagstrum and Becker 1973). The neutralization step from $He^{2+} \rightarrow H^+$ and subsequent electron emission can be represented as

$$M + He^{2+} \rightarrow M^{2+} + H^+ + e^- \tag{4.1}$$

In this process, the metal M loses two electrons. One changes the charge state of the ion and one is emitted. Then the highest energy of the emitted electron is the ionization energy 54.42 eV minus twice the work function. The work function for the crystal Ir(100) is 5.76 eV; therefore an electron emitted from this surface has a maximum energy of 42.9 eV. In analogy, one can write for the neutralization step from $He^+ \rightarrow He^0$:

$$M + He^+ \rightarrow M^{2+} + He^0 + e^- \tag{4.2}$$

Again the surface provides two electrons, one of which is ejected into the vacuum. The maximum energy of this electron is the ionization energy 24.59 eV minus twice the work function of Ir(100), which yields 13.07 eV. After the neutralization of He^{2+} $\rightarrow He^0$, the energy sum of the two emitted electrons has an upper bound defined as $E_{\text{max sum}} = DI - 4 \times e\varphi$; the factor 4 is a consequence of the fact that the metal provides four electrons that leave the surface. The numerical value of $E_{\text{max sum}}$ is 55.97 eV for Ir(100). It is possible that the neutralization does not yield the He atom in the ground state, but in its first excited state (19.82 eV above the ground state). For the low ion energies used in experiment (Li et al. 2017), this pathway can be ignored (Hagstrum 1966). The question arises as to whether a single He^{2+} ion can lead to electron pair emission. This can only be answered by a coincidence experiment.

For detection of correlated electron pairs emitted from an Ir(001) surface induced by He^{2+} ions the same coincidence spectrometer was used as for the (e,2e) measurements on Cu(111) (Schumann et al. 2011). It incorporates two hemispherical electron energy analysers. They are labelled as "left" and "right". The two transfer lenses include an angle of $90°$ and define the reaction plane, which comprises the normal to the surface. The "normal to the surface 10 eV incident He^{2+} beam" was delivered by an ion source and a Wien mass filter (Tusche and Kirschner 2014). The emitted electrons are detected with energies E_{left} and E_{right}. To check the existence of time-correlated electron pairs originating from a single ion –surface collision the arrival time difference $\Delta t = t_{\text{left}} - t_{\text{right}}$ distribution was measured. Here Δt is the time difference between detection of "left" and "right" electrons. Electron pair emission upon single-particle excitation exists if such distribution displays a peak provided the incident ion current is sufficiently small (Schumann et al. 2011). Such a peak shows up in Fig. 4.9 for a Ir(100) surface excited by a 10 eV He^{2+} beam. This is the first direct proof that the neutralization of one He^{2+} ion leads to electron pair emission (true coincidences). The peak width reflects the time resolution of the instrument (Kugeler et al. 2003). The constant background (random coincidences) is due to electron emission caused by two He^{2+} ions. The observation of pair emission due to He^{2+} neutralization is not a trivial point. Using 5 eV He^+ ions shows no evidence of pair emission, see inset of Fig. 4.9, while the energy gain in this neutralization (24.59 eV) would be sufficient for pair emission. Analysis of the two-dimensional energy distribution of correlated electron pairs helps to get insight into the mechanism of the pairs' generation. Each of the spectrometers covered an energy window of ± 13.5 eV around a central energy E_k. The energy resolution of the spectrometers was 0.7 eV. Two coincidence spectra were measured: one with setting of $E_k = 19$ eV

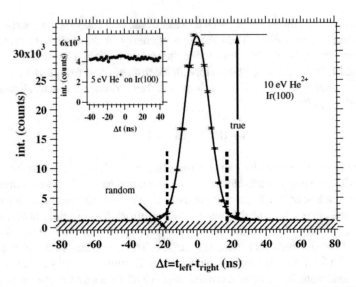

Fig. 4.9 Arrival time histogram of electron pairs emitted from Ir(100) due to 10 eV He^{2+} ions. The peak is indicative of "true" coincidences while the constant background arises from "random" coincidences. The dashed vertical lines indicate the selection of events for further analysis. Excitation with 5 eV He$^+$ ions showed no evidence of pair emission; see inset. (Reprinted figure with permission from Li et al. (2017), Copyright (2017) by APS)

and the second with E_k = 30 eV in both spectrometers. The 2D-energy spectra upon He^{2+} excitation are shown in Fig. 4.10. Events within the interval marked by the dashed vertical lines in Fig. 4.9 were selected (Li et al. 2017).

Within the L-shaped region (about 10 eV of width along both energy axes) one electron can have energy up to 42.9 eV while the other cannot exceed 13.07 eV. Therefore, this region contains events in which one electron is emitted first upon neutralization from He^{2+} → He$^+$ while the second is due to the neutralization from He$^+$ → He^0.

Since 98% of events are located in the L-shaped region of the two-dimensional energy distribution of pairs (Fig. 4.10) the two neutralization steps leading to He^0 proceed mainly independently (Li et al. 2017). The dashed diagonal lines in both panels indicate the position of $E_{\text{max sum}}$ = 55.97 eV of pairs.

A small but sizable intensity resides outside the L-shaped region, which suggests an additional pathway of pair emission. To highlight this part of the correlated pairs distribution the central energy of spectrometers was shifted to E_k = 30 eV and the L-shaped region is outside the field of view (see Fig. 4.10b) so that the spectrum now contains events that cannot be explained by a sequence of neutralization steps. The intensity is highest if both electrons have an energy at the lower part of the detection window. The intensity gradually decreases if one moves closer to the dashed diagonal line marking $E_{\text{max sum}}$. However, a cut-off value where the intensity drops sharply cannot be identified on the spectrum. For a more detailed view, the energy sum

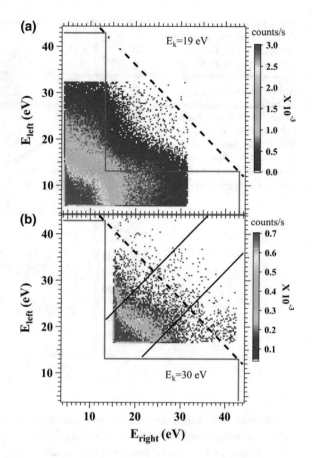

Fig. 4.10 2D-energy spectra for a central energy of $E_k =$ 19 eV (**a**) and $E_k = 30$ eV (**b**). The red lines mark an L-shaped region; events within this originate from a sequential emission. The dashed black diagonal lines mark $E_{max\ sum}$. The pair of black solid lines in (**b**) label the region $(E_{right} - E_{left}) \leq$ 8 eV. (Reprinted figure with permission from Li et al. (2017), Copyright (2017) by APS)

$(E_{left} + E_{right})$ distribution is plotted. Furthermore, for this histogram the events for which both electrons have similar energies were chosen. This part of pairs' distribution is not covered by a scenario of sequential emission. Figure 4.11a presents the E_{sum} energy spectrum obtained from the data set plotted in Fig. N(b) under the constraint $(E_{right} - E_{left}) \leq 8$ eV. Note that the wide band shown by solid diagonal lines in Fig. 4.10b is used to have reasonable statistics.

The E_{sum}-spectrum shows a monotonic decrease without a sharp cut-off near $E_{max\ sum}$. This energy position is marked by the dashed vertical line. However, the Δt histograms for pairs above and below this line are different: Δt histogram for pairs below $E_{max\ sum}$ line shows the maximum (meaning existence of correlated pairs), but such a histogram for pairs above this line does not show any maximum (meaning the contribution of accidental coincidences of electrons generated by different ions). Figure 4.11 indicates that there is electron pair emission associated with neutralisation of a single He^{2+} ion into a He atom in the ground state. This means the full double ionization energy is available for electron pair emission. It can be concluded that a sequence of neutralization steps does not account for the coincidence events observed in Fig. 4.10b (Li et al. 2017).

Fig. 4.11 Panel (**a**) shows the E_{sum} spectrum of the data presented in Fig. 4.10 (**b**) under the constraint ($E_{right} - E_{left}$) ≤ 8 eV. The vertical dashed line marks $E_{max\ sum}$. In (**b**) the Δt histogram is shown for an E_{sum} range just below or above $E_{max\ sum}$ as indicated by the two boxes in (**a**). (Reprinted figure with permission from Li et al. (2017), Copyright (2017) by APS)

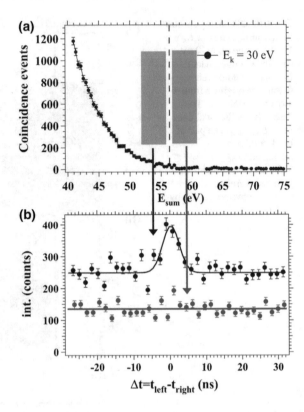

As above mentioned only 2% of coincidence events are located in the L-shaped area of two-dimensional energy distribution of pairs in Fig. 4.10a. With this information a simple picture can be proposed to estimate the time scale τ for the correlated electron emission (Li et al. 2017). The two neutralization steps occur on average within a time t_{av} and it is assumed that they proceed independently. The small intensity contribution outside the L-shaped region indicates that $t_{av} \gg \tau$. The probability for the two neutralization steps to occur within a time interval τ is then given by the Poisson distribution $P(\tau/t_{av}) \approx \tau/t_{av}$. If the intensity contribution outside the L-shaped area is a measure of t_{av}/τ, one obtains $\tau = 0.02 t_{av}$. One can adopt for t_{av} values in the range 2–20 fs on the basis of neutralization rates (Lorente and Monreal 1994; Wethekam et al. 2003). This finally yields $\tau = 40-400$ as. Hence, two formally independent neutralization steps occurring within an interval shorter than τ are recognized as a single excitation for electron pair emission.

Drawing analogy with atomic cases the following microscopic picture can be envisaged (Li et al. 2017). A core vacancy of the atom is created and then filled by an electron from an outer shell, the available energy is transferred to two electrons (see Fig. 4.12a). This double Auger decay has been observed in coincidence spectroscopy from Ar (Viefhaus et al. 2004). It was also discovered that the electron emission spectrum excited from Ni surface and a carbon foil by multiply charged C^{n+} and N^{m+} ions contained intensity at twice the nominal Auger electron energy of the projectile

Fig. 4.12 In (**a**) a core hole is filled by an electron and pair emission called double Auger decay takes place (Viefhaus et al. 2004). In the three electron Auger (**b**), a double vacancy is filled by an electron pair and the gained energy is transferred to a third electron (Folkerts et al. 1992; De Filippo et al. 2008). In (**c**) an electron pair of the metal fills the double vacancy and an electron pair from the metal is emitted. (Reprinted figure with permission from Li et al. (2017), Copyright (2017) by APS)

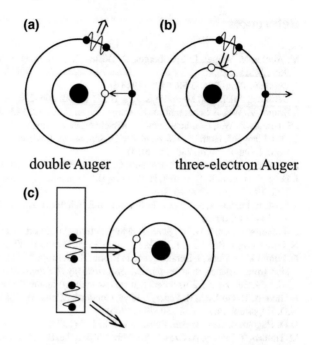

(Folkerts et al. 1992; De Filippo et al. 2008). This result was explained by a three-electron Auger decay (see Fig. 4.12b). A double core hole is filled by an electron pair, and this energy is transferred to another single electron. Data (Li et al. 2017) are interpreted by a combination of the double Auger decay and a three-electron Auger decay (see Fig. 4.12c). Upon approaching the surface the double vacancy of the He^{2+} ion is filled by two electrons similar to the three-electron Auger process. The key difference is that these electrons do not originate from higher lying orbitals of the atom, but come from the metal surface. The energy gain is transferred to an electron pair of the surface which is emitted. This view is further corroborated by a recent work in which the electron capture for He^{2+} ions into excited states proceeds in a single step (Tusche 2015). Extending this picture of a correlated double electron capture, a single step of an electron pair from the metal into the ground state of the He atom was observed (Li et al. 2017). If a single electron were to gain the energy of the two neutralization steps, it would have energy up to 62 eV, but this is not observed and is consistent with the explanation of two formal neutralization steps by the pair emission. The neutralization of a single He^{2+} ion near a metal surface leads to electron pair emission. A sizable number of these pairs originate from a correlated single step neutralization of the ion involving a total of four electrons from the metal. On the time scale of the correlated electron dynamics, these two neutralization steps do not proceed sequentially. The time scale of the correlated electron dynamics is estimated to be in the range of 40–400 as.

References

M. Achler, V. Mergel, L. Spielberger, R. Dörner, Y. Azuma, H. Schmidt-Böcking, J. Phys. B **34**, 965 (2001)

J. Berakdar Phys. Rev. B **58**, 9808 (1998-I)

J. Berakdar, Nucl. Inst. Meth. Phys. Res. B **171** (2000) 204–218

I. Brandt, Z. Wei, F. Schumann, J. Kirschner, Phys. Rev. Lett. **113**, 107601 (2014)

I.S. Brandt, Z. Wei, F.O. Schumann, J. Kirschner, Phys. Rev. B **92**, 155106 (2015)

C.-T. Chiang, M. Huth, A. Trützschler, F.O. Schumann, J. Kirschner, W. Widdra, J. Electron Spectrosc. Relat. Phenom. **200**, 15 (2015)

E. De Filippo, G. Lanzanò, H. Rothard, C. Volant, Phys. Rev. Lett. **100**, 233202 (2008)

J. Feist, S. Nagele, R. Pazourek, E. Persson, B.I. Schneider, L.A. Collins, J. Burgdörfer, Phys. Rev. Lett. **103**, 063002 (2009a)

J. Feist, R. Pazourek, S. Nagele, E. Persson, B.I. Schneider, L.A. Collins, J. Burgdörfer, J. Phys. B **42**, 134014 (2009b)

L. Folkerts, J. Das, S.W. Bergsma, R. Morgenstern, Phys. Lett. A **163**, 73 (1992)

N. Fominykh, J. Berakdar, J. Henk, P. Bruno, Phys. Rev. Lett. **89**, 086402 (2002)

N. Fominikh, J. Henk, J. Berakdar, and P. Bruno, H. Gollisch, R. Feder, Double-electron photoemission from surfaces, in *Mani-particle Spectroscopy of Atoms, Molecules, Clusters and Surfaces*, ed. by J. Berakdar, J. Kirschner (Kluwer, Academic/Plenum Publishers, New York, 2001), p. 461

F. Giebels, H. Gollisch, R. Feder, J. Phys, Condens. Matter **21**, 355002 (2009)

H.D. Hagstrum, Phys. Rev. **150**, 495 (1966)

H.D. Hagstrum, G.E. Becker, Phys. Rev. B **8**, 107 (1973)

M. Hattass, T. Jalowy, A. Czasch, Th Weber, T. Jahnke, S. Schössler, LPh Schmidt, O. Jagutzki, R. Dörner, H. Schmidt-Böcking, Rev. Sci. Instrum. **75**, 2373 (2004)

M. Hattass, T. Jahnke, S. Schössler, A. Czasch, M. Schöffler, LPhH Schmidt, B. Ulrich, O. Jagutzki, F.O. Schumann, C. Winkler, J. Kirschner, R. Dörner, H. Schmidt-Böcking, Phys. Rev. B **77**, 165432 (2008)

R. Herrmann, S. Samarin, H. Schwabe, J. Kirschner, PRL **81**, 2148 (1998)

F.U. Hillebrecht, A. Morozov, J. Kirschner, Phys. Rev. B **71**, 125406 (2005)

B.W. Hoogenboom, R. Hesper, L.H. Tjeng, G.A. Sawatzky, Phys. Rev. B **57**, 11 939 (1998)

M. Huth, C.-T. Chiang, A. Trützschler, F.O. Schumann, J. Kirschner, W. Widdra, Appl. Phys. Lett. **104**, 061602 (2014)

O. Kugeler, S. Marburger, U. Hergenhahn, Rev. Sci. Instrum. **74**, 3955 (2003)

C.H. Li, C. Tusche, F.O. Schumann, J. Kirschner, PRL **118**, 136402 (2017)

R.W. Lof, M.A. van Veenendaal, B. Koopmans, H.T. Jonkman, G.A. Sawatzky, Phys. Rev. Lett. **68**, 3924 (1992)

N. Lorente, R. Monreal, Surf. Sci. **303**, 253 (1994)

M.L.E. Oliphant, Proc. R. Soc. A **127**, 373 (1930)

F.O. Schumann, C. Winkler, G. Kerherve, J. Kirschner, Phys. Rev. B **73**, 041404(R) (2006)

F.O. Schumann, C. Winkler, J. Kirschner, PRL **98**, 257604 (2007)

F.O. Schumann, R.S. Dhaka, G.A. van Riessen, Z. Wei, J. Kirschner, Phys. Rev. B **84**, 125106 (2011)

T. Andreas, M. Huth, C.-T. Chiang, R. Kamrla, F.O. Schumann, J. Kirschner, W. Widdra, Phys. Rev. Let. **118**, 136401 (2017)

K.-D. Tsuei, J.Y. Yuh, C.-T. Tzeng, R.-Y. Chu, S.-C. Chung, K.-L. Tsang, Phys. Rev. B **56**, 15 412 (1997)

C. Tusche, J. Kirschner, Rev. Sci. Instrum. **85**, 063305 (2014)

C. Tusche, Phys. Rev. Lett. **115**, 027602 (2015)

G.A. van Riessen, F.O. Schumann, M. Birke, C. Winkler, J. Kirschner, J. Phys.: Conf. Ser. **185** (2009) 012051

G.A. van Riessen, F.O. Schumann, M. Birke, C. Winkler, J. Kirschner, J. Phys, Condens. Matter **20**, 442001 (2008)

J. Viefhaus, S. Cvejanović, B. Langer, T. Lischke, G. Prümper, D. Rolles, A.V. Golovin, A.N. Grum-Grzhimailo, N.M. Kabachnik, U. Becker, Phys. Rev. Lett. **92**, 083001 (2004)
S. Wethekam, A. Mertens, H. Winter, Phys. Rev. Lett. **90**, 037602 (2003)

Epilogue

The use of low-energy spin-polarized electrons in a beam has highlighted the development of studying electron-electron interactions, electron correlations and the mechanisms of electron emission and electronic properties of surfaces. Most, if not all, of our experimental explorations have been presented in a way which demonstrates the experimental approach of how and why the sequence of steps has been made. We hope readers have been influenced in at least two ways. First, that they have been encouraged to explore quantum phenomena at the fundamental levels of symmetry, of orbital and spin angular momenta in the reaches of expanding space and time, hopefully in their own laboratory. Secondly we hope that readers have increased appreciation of the vast expanse of essential supporting knowledge in books indicated in the Preface. The collaborating theoretical knowledge of many researchers has helped our understanding of the quantum expanse underpinning the experiments and the vast many-body nature of correlation spectroscopy.

Underpinning all aspects of this book and beyond, but particularly with focal points in each word of the title of the book, are the extent of single incident-electron scattering events and two-particle correlation detection. What are the quantum definition and expectation for each observable? It was the choice of instruments, their settings and their use that determined the momentum and time (energy) of detection and hence the particle correlations. All these features are expected to be emphasized, and their limits extended, in future studies. However, observations of two electrons emerging from a surface (and/or its constituent atoms or molecules) under a single electron impact shows they are well correlated.

Those thoughts are evolving with the present interpretations of modern quantum mechanics. For electron scattering by another electron the concept of their spatial and temporal interaction expands because of their Coulomb charge and its range. The concept of non-locality is satisfying measurements and ideas of "quantum entanglement".

If the basic concept of a single atom include an infinity of virtual harmonic oscillators how is the spatial and temporal development and observation of double

S. Samarin et al., *Spin-Polarized Two-Electron Spectroscopy of Surfaces*, Springer Series in Surface Sciences 67, https://doi.org/10.1007/978-3-030-00657-0

and multiple photon observation explained? How are the ideas developed for single and multiple electron emission from an atom, molecule or surface of a metal, semiconductor or insulator, and perhaps in coincidence with one of more photons? What are the geometric or topological paths for all particles as they follow some 'impact particle perturbed' potential until "free"? What are the smallest spatial and time scales to observe a spin exchange, or a spin-orbit interaction? How do the quantum mechanisms and observable characterizations change from a localized atom to an extended thin film on a magnetic surface?

Nevertheless, within the present experimental limits of some pico- and even atto-seconds with photon assistance, our experiments "observe' that imagined events acquire reality and allow multiple correlations in space and time. It is hoped that readers will question our observations and seek further experimental and theoretical evidence and interpretations and contribute to the knowledge and applications of quantum physics.

The authors: Perth and St. Petersburg. February 2018.

Index

© Springer Nature Switzerland AG 2018
S. Samarin et al., *Spin-Polarized Two-Electron Spectroscopy of Surfaces*, Springer
Series in Surface Sciences 67, https://doi.org/10.1007/978-3-030-00657-0

Printed in the United States
By Bookmasters